Advances in Intelligent Systems and Computing

Volume 724

Series editor

Janusz Kacprzyk, Polish Academy of Sciences, Warsaw, Poland
e-mail: kacprzyk@ibspan.waw.pl

About this Series

The series "Advances in Intelligent Systems and Computing" contains publications on theory, applications, and design methods of Intelligent Systems and Intelligent Computing. Virtually all disciplines such as engineering, natural sciences, computer and information science, ICT, economics, business, e-commerce, environment, healthcare, life science are covered. The list of topics spans all the areas of modern intelligent systems and computing.

The publications within "Advances in Intelligent Systems and Computing" are primarily textbooks and proceedings of important conferences, symposia and congresses. They cover significant recent developments in the field, both of a foundational and applicable character. An important characteristic feature of the series is the short publication time and world-wide distribution. This permits a rapid and broad dissemination of research results.

Advisory Board

Chairman

Nikhil R. Pal, Indian Statistical Institute, Kolkata, India
e-mail: nikhil@isical.ac.in

Members

Rafael Bello Perez, Universidad Central "Marta Abreu" de Las Villas, Santa Clara, Cuba
e-mail: rbellop@uclv.edu.cu

Emilio S. Corchado, University of Salamanca, Salamanca, Spain
e-mail: escorchado@usal.es

Hani Hagras, University of Essex, Colchester, UK
e-mail: hani@essex.ac.uk

László T. Kóczy, Széchenyi István University, Győr, Hungary
e-mail: koczy@sze.hu

Vladik Kreinovich, University of Texas at El Paso, El Paso, USA
e-mail: vladik@utep.edu

Chin-Teng Lin, National Chiao Tung University, Hsinchu, Taiwan
e-mail: ctlin@mail.nctu.edu.tw

Jie Lu, University of Technology, Sydney, Australia
e-mail: Jie.Lu@uts.edu.au

Patricia Melin, Tijuana Institute of Technology, Tijuana, Mexico
e-mail: epmelin@hafsamx.org

Nadia Nedjah, State University of Rio de Janeiro, Rio de Janeiro, Brazil
e-mail: nadia@eng.uerj.br

Ngoc Thanh Nguyen, Wroclaw University of Technology, Wroclaw, Poland
e-mail: Ngoc-Thanh.Nguyen@pwr.edu.pl

Jun Wang, The Chinese University of Hong Kong, Shatin, Hong Kong
e-mail: jwang@mae.cuhk.edu.hk

More information about this series at http://www.springer.com/series/11156

Tatiana Antipova · Álvaro Rocha
Editors

Information Technology Science

 Springer

Editors
Tatiana Antipova
The Institute of Certified Specialists
Perm
Russia

Álvaro Rocha
Universidade de Coimbra
Coimbra
Portugal

ISSN 2194-5357   ISSN 2194-5365   (electronic)
Advances in Intelligent Systems and Computing
ISBN 978-3-319-74979-2   ISBN 978-3-319-74980-8   (eBook)
https://doi.org/10.1007/978-3-319-74980-8

Library of Congress Control Number: 2017964591

Printed on acid-free paper

This Springer imprint is published by the registered company Springer International Publishing AG part of Springer Nature
The registered company address is: Gewerbestrasse 11, 6330 Cham, Switzerland

Preface

This book contains a selection of papers accepted for presentation and discussion at The 2017 International Conference on Information Technology Science (MosITS'17). This Conference had the support of the Institute of Certified Specialists, Russia, AISTI (Iberian Association for Information Systems and Technologies), and Springer. It took place at Izmailovo Convention Centre, Moscow, Russia, on December 01–03, 2017.

MosITS'17 is an international forum for researchers and practitioners to present and discuss the most recent innovations, trends, results, experiences, and concerns in the several perspectives of Information Technology Science.

The Scientific Committee of MosITS'17 was composed of a multidisciplinary group of 60 experts and those who are intimately concerned with Information Technology in Science. They have had the responsibility for evaluating, in a "double-blind review" process, the papers received for each of the main themes proposed for the Conference: Information Technology in Communication, Information Technology in Management Science, Information Technology in Public Administration, Information Technology in Economics, Information Technology in Business & Finance, Information Technology in History, Information Technology in Health & Rehabilitation, Information Technology in Education, Information Technology in Art.

MosITS'17 received about 40 contributions from 16 countries around the world. The papers accepted for presentation and discussion at the Conference are published by Springer (this book) and will be submitted for indexing by ISI, SCOPUS, among others.

We acknowledge all of those that contributed to the staging of MosITS'17 (authors, committees, workshop organizers, and sponsors). We deeply appreciate their involvement and support that was crucial for the success of MosITS'17.

November 2017

Tatiana Antipova
Álvaro Rocha

Organization

Scientific Committee

Tatiana Antipova (Chair)	Institute of Certified Specialists, Russia
Álvaro Rocha (Co-chair)	University of Coimbra, Portugal
Alexandru Stancu	University of Geneva, Switzerland
Anatoli Bourmistrov	Nord University Business School, Norway
Andre Aquino	University of São Paulo, Brazil
António José Abreu Silva	ISCAP, Portugal
Elena Dybtsyna	Nord University Business School, Norway
Filipe Alexandre Almeida Ningre de Sá	Municipality of Penacova, Portugal
Giuseppe Grossi	Kristianstad University, Sweden
Indra Bastian	University Gadjah Mada, Yogyakarta, Indonesia
Isabel Pedrosa	University of Coimbra, Portugal
Joao Vidal de Carvalho	Polytechnic Institute of Porto, Portugal
Karl Gratzer	Sodertorn University, Sweden
Linda Kidwell	Nova Southeastern University, Fort Lauderdale, USA
Luca Bartocci	University of Perugia, Italy
Mikhail Kuter	Kuban State University, Russia
Nikolai Ustiuzhantsev	Perm State Medical University, Russia
Oleg Golosov	"Fors" Group of companies, Moscow, Russia
Seppo Sirkemaa	University of Turku, Finland
Simona Riurean	University of Petrosani, Romania
Suzanne Lowensohn	University of Vermont, Burlington, USA
Tjerk Budding	Vrije Universiteit Amsterdam, Amsterdam, Netherland

Contents

Information Technology in Communication

Information Technology in Health and Rehabilitation

eHealth Infrastructures and Security in Portuguese Hospitals: Benchmarking with European Hospitals

João Vidal Carvalho[1]([⊠]) [iD], Álvaro Rocha[2], and António Abreu[1]

[1] Politécnico do Porto, ISCAP, CEOS.PP, São Mamede de Infesta, Portugal
{cajvidal, aabreu}@iscap.ipp.pt
[2] Departamento de Engenharia Informática, Universidade de Coimbra,
Coimbra, Portugal
amrocha@dei.uc.pt

Abstract. The aim of this study is to describe and compare eHealth Infrastructures and Security in selected European Union (EU) countries. Using data available from the European Hospital Survey (2012/2013), the study reported in this paper analyses the adoption of five eHealth Infrastructures and Security indicators in Portuguese Hospitals, namely: Externally connected; Broadband; Single and unified wireless; Clear and structured rules on access to clinical data; and EAS (Enterprise Archiving Strategy) for disaster recovery in less than 24 h. The analysis of these eHealth indicators allows us to understand the position of Portugal in relation to the European average. It has been found that the Portuguese Hospitals are well positioned with regard to the adoption of these indicators and in spite of this, remains a permanent concern to raise the eHealth Infrastructures and Security levels through new projects in progress.

Keywords: eHealth · Hospital Information Systems
Health information systems · eHealth Infrastructures · eHealth Security
Portugal

1 Introduction

The rapid development of the Information and Knowledge Society and consequently, the rapid advancement of Information and Communication Technologies (ICT) have revolutionized the way we interact with each other. The convergence between the acceleration capabilities of computers, the range and expansion of the Internet and the increase in the ability to capture and leverage the knowledge in digital format are key drivers for the technological revolution that we live nowadays. The current information society, has the potential to cause a revolution similar in health care [1]. It could change the relationship between the patient and the professional, providing valuable opportunities for health professionals rendering healthcare services effectively through the use of Information Systems and Technologies (IST) to their patients and providing them ease access to relevant (clinical) information. However, healthcare systems around the world, currently, are facing considerable pressure to reduce costs, enhance and improve service efficiency, expand access while maintaining or even improving the quality of

© Springer International Publishing AG, part of Springer Nature 2018
T. Antipova and Á. Rocha (Eds.): MosITS 2017, AISC 724, pp. 3–13, 2018.
https://doi.org/10.1007/978-3-319-74980-8_1

health services provided [2–4]. The side effects, such as demographic trends, the lack of qualified health professionals and the expectations and demands of patients, local administrators or health insurers come to hinder the fulfilment of this mission [5]. There are strong expectations that a wider adoption of IST in the health field will contribute to improve the health of individuals and the performance of providers, yielding improved quality, cost savings, and greater engagement by patients in their own healthcare [6]. However, there is evidence that the implementation of the IST without the adaptation of the infrastructures and the strategic/organizational processes behind it, will not necessarily generate the expected benefits [7]. Health institutions, in line with government organizations, are starting to realize that the reasons behind a certain inability to properly manage health processes are directly related to the limitations of technological infrastructures and lack of efficiency in their management [8, 9].

In this context, appears the concept of eHealth defined by Sharma [9] as: *"The way to achieve the best results of the health process through the effective and innovative use of Information Technologies (IT)"*. eHealth is described as the application of IST Infrastructure across the spectrum of functions that affect the health sector [10]. eHealth includes tools for health authorities and their professionals, as well as, customized health systems for patients and citizens. eHealth may therefore be adopted to cover the interaction between patients and health care providers, may include health information networks, electronic health records, telemedicine services and personal portable systems, in support of management of prevention, diagnosis, treatment, follow-up of patients' health and lifestyle.

eHealth also offers significant opportunities for health professionals to provide technologically effective services to their consumers by providing them with ways to access the information they need. In the same sense, the European Commission (EC) clarifies that eHealth is *"the use of ICT in health products, services and processes combined with organizational change in healthcare systems and new skills, in order to improve health of citizens, efficiency and productivity in healthcare delivery, and the economic and social value of health"* [11]. From the standpoint of this study: *"eHealth covers the interaction between patients and health-service providers, institution-to-institution transmission of data, or peer-to-peer communication between patients and/or health professionals"* [11].

In this article, we will initially present the indicators of a study carried out in Europe in 2012/2013 that will be analyzed and compared with Portugal Healthcare reality. Subsequently, the study carried out by the EU will be described. After that, the values of five eHealth Infrastructures and Security indicators will be compared between Portugal and the European average. Finally, we will present recent projects carried out in Portugal that suggest a positive evolution of the indicators considered in this study.

2   eHealth Infrastructures and Security Indicators

Based on the eHealth Benchmarking III study from the European Hospital Survey (2012/2013) [12] (described in Sect. 3), country profiles have been built using 13 eHealth indicators. These eHealth indicators encompass Infrastructures, Applications,

Integration and Security and consist in specific answers to the questionnaire and identify the eHealth best practices in Europe.

Regarding the study presented in this article, a selection of five Infrastructures and Security items was made. In the case of infrastructures, the following indicators are used: Externally connected; Broadband > 50 Mbps; Single and unified wireless. In the case of Security, the following two indicators are used: Clear and structured rules on access to clinical data; EAS for disaster recovery in less than 24 h.

2.1   Infrastructure

As in other activity sectors, hospitals must resort to IST and their infrastructure to support all their activities, inside the hospital environment and where different health field partners are involved [13]. In the same alignment, Sharma [9] refers that the system connected with the healthcare process can be defined as: "*a set of activities, methods and practices that people use to provide health services and maintain the environment that supports service providers*". This environment involves both the medical devices, as well as the healthcare entities associated with the supply and, fundamentally, the IT infrastructure. Regarding the study presented in this article, a selection of three eHealth Infrastructures indicators are used to identify pre-requisites towards "ubiquitous eHealth systems". These systems enable remote patient monitoring and health information exchange beyond hospital boundaries [12]. The three indicators used are:

- *Externally connected*: importance of access to the infrastructure outside the hospital-specific site. Extranet systems, value-added networks and proprietary infrastructures enable inter-connectivity between healthcare stakeholders and hence ensure a high level of health care;
- *Broadband > 50 Mbps*: a high-speed broadband was one of the most important policy priorities within the EC's Digital Agenda for Europe. Should allow the processing and transfer of an increasing amount of data, such as images, reports and telemonitoring services;
- *Single and unified wireless*: such infrastructure should allow mobile access to different applications and services in every location of the hospital.

2.2   Security

The main goals of data security are confidentiality, integrity and availability. However, safeguarding these goals does not translate automatically into security measures for health organizations. Security is achieved by simultaneously preventing attacks against IST and guaranteeing that the mission of the organization is fulfilled, despite the attacks and accidents [14]. Indeed, security issues in organizations stem from the fact that security is many times addressed individually and without any connection with the business goals. In Hospital Information Systems, ensuring security and privacy of data is required to build trust between the medical staff, the patient, and the other stakeholders who may need patient clinical information [12]. If risks are perceived or information is not accurate or partially complete, patients or physicians may not be willing to disclose necessary health information, which could be life-threatening [12]. Two indicators have been selected to reflect the level of security in acute care hospitals:

- *Clear and structured rules on access to clinical data*: to ensure privacy of data, access to certain types of data must be restricted to some specific healthcare professionals;
- *EAS for disaster recovery in less than 24 h*: an EAS enables users to restore clinical information facilities and information when necessary. This indicator reflects the hospital robustness to provide the services to ensure continuity of performance.

3 eHealth Benchmark Study in Europe

The European Union, through its executive body, the European Commission (EC), has been a very active stakeholder in promoting the digital agenda in health in recent decades [15, 16]. Since 1989, the EC has invested over €1 billion in over 450 eHealth projects [16]. The work includes action plans for eHealth (e.g. [11]) directives and guidelines related to eHealth (e.g. [17]), sponsored eHealth projects (e.g. [18]), benchmarking activities (e.g. [19]) and commissioned research (e.g. [12]).

The EC has undertaken extensive research into the adoption of eHealth in European countries in the last decade, and this material was used in the study. The EC commissioned three major surveys into the adoption of eHealth in primary care in 2002 [19], 2007 [20] and 2013 [12] in 15, 29 and 30 countries, respectively. In the last survey (2013), the total sample are 1753 Hospitals of which 41 are Portuguese. Taken together, this qualitative and quantitative research paints a rich and complex picture of the development and adoption of eHealth in Europe over the past two decades.

The 2013 EC survey studied general practitioners (GP) in the primary care setting, which the report defined as "physicians working in outpatient establishments in specialties such as general practice, family doctor, internal medicine and general medicine" [12]. The survey study took 18 months, and 9196 GP (2%) from 30 European countries were surveyed in detail about the adoption of eHealth in primary care.

3.1 Limitations of the Study

This study was limited by the availability of comparable data for 30 countries and by material in the English language. Other European countries in the Balkan area, Eastern Europe and Switzerland were not included in the study. This study focuses mainly on aspects of eHealth, with a comparatively higher level of detail. However, questions of general scope such as hospital growth, financial position or strategic direction are missing. As a result of missing information for some explanatory variables, the analysis is based on smaller samples. It is probable that some countries have made further progress since their data were published in 2014, and the adoption rates in these countries may be higher than those reported in this study.

3.2   Questions Used in the Study for Infrastructures and Security Indicators

To build the eHealth indicators, we have relied on the most important questions of the survey. Table 1 presents these questions and answer options used to define the eHealth Infrastructures and Security indicators and assess the implementation of eHealth in European acute[1] care hospitals.

Table 1. Question items used for the eHealth indicators.

eHealth indicator	Question used	Answer option used
Externally connected	Is your hospital computer system externally connected...?	At least one of the two following answers: - Yes, through an extranet i.e. using a secure internet connection over the internet - Yes, through a value-added network or proprietary infrastructure
Broadband > 50 Mbps	What type of internet connection does your hospital have?	- Broadband (from 50 Mbps to 100 Mbps) - Broadband (above 100 Mbps)
Single and unified wireless	How does your hospital support wireless communications?	There is a single, unified wireless infrastructure capable of supporting most of the applications
Clear and structured rules on access to clinical data	Are there clear structured rules on accessing (reading-writing) patients' electronic medical data?	Yes
EAS for disaster recovery in less than 24 h	Please estimate how quickly your organization can restore critical clinical information system operations if a disaster causes the complete loss of data at your hospital's primary data centre	- Immediate (we have a fully redundant data centre) - Less than 24 h

3.3   Portugal's Acute Hospitals EHealth Profile

The study identified 589 hospitals in Portugal. Within this sample, 224 (38%) completed the screener part of the questionnaire and, of these, 12% qualified as acute care hospitals. Of the 73 screened in, 41 acute hospitals (56%) completed the survey, mostly under 250 beds (Table 2).

---

[1] The study defined the following criteria to qualify survey participants as acute care hospital: The hospital has an emergency department, and at least a routine and/or life-saving surgery operating room and/or an intensive care unit.

Table 2. Portuguese breakdown by size of hospital.

	Hospitals	<101 beds	≥ 101 and ≤ 250	≥ 251 and ≤ 750	>750 beds	No answer
Census	73	21 (29%)	16 (22%)	7 (10%)	3 (4%)	26 (36%)
Survey2013	41	13 (32%)	11 (27%)	6 (15%)	3 (7%)	8 (20%)

Table 3 shows that most hospitals involved in the study are public.

Table 3. Portuguese breakdown by ownership type.

	Hospitals	Public	Private	Private (not profit)	No answer
Census	73	34 (47%)	20 (27%)	8 (11%)	11 (15%)
Survey2013	41	24 (59%)	13 (32%)	4 (10%)	

3.4   Position of the Portuguese eHealth Profile in EU27 + 3

The following figure presents the spider chart showing the profile of the Portuguese Hospitals in the indicators evaluated in the 2013 study. It is observed in Fig. 1 that the score presented in the infrastructure items show values higher than the European average, especially Broadband which is substantially higher than the European Hospitals average. On the other hand, one of the eHealth Security indicator (EAS for disaster recovery in less than 24 h) has values below the average of European hospitals and the other (Clear and structured rules on access to clinical data) is in line with the average.

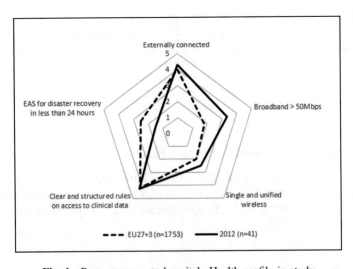

Fig. 1. Portuguese acute hospital eHealth profile in study

Table 4 shows that Portugal is close to the European average in its eHealth profile. However, the gains over and above the European average are not evenly distributed. While the infrastructure indicators show values above the European average, as is the case for "Broadband > 50 Mbps" with 31% above the EU27 + 3 average. Similarly, "Single and unified wireless" was 10% and "Externally connected" was 5% above the average. On the other hand, the safety indicators are below the EU27 + 3 average.

Table 4. Portuguese eHealth indicators[2]

eHealth indicator	Score difference Portugal vs EU27 + 3
Externally connected	5%
Broadband > 50 Mbps	31%
Single and unified wireless	10%
Clear and structured rules on access to clinical data	−1%
EAS for disaster recovery in less than 24 h	−17%

Private acute hospitals in Portugal appear to be the best endowed in terms of eHealth security capabilities, while Public hospitals leading in Infrastructures. However, it is important to note that the indicators analyzed in the study were limited to "Broadband > 50 Mbps" and "Clear and structured rules on access to clinical data" (Fig. 2).

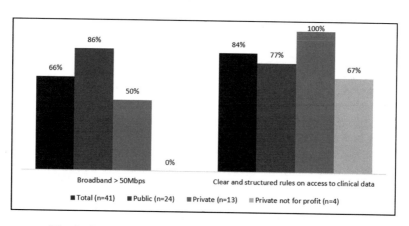

Fig. 2. Portuguese acute hospitals eHealth profile by ownership

---

[2] Note: results are based on valid answers only. The scoring scale from 0 to 5 points corresponds to an implementation rate from 0% to 100%.

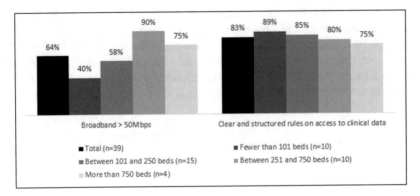

Fig. 3. Portuguese acute hospitals eHealth profile by size

The distribution of eHealth Security indicator (Clear and structured rules on access to clinical data) appears to be relatively even in terms of hospital size. (Figure 3). Relatively to eHealth capability (Broadband > 50 Mbps), large hospitals have higher values than smaller hospitals (Fig. 3).

4   Portuguese eHealth Infrastructures/Security Projects in Progress

After the publication of these results in 2014, it is very likely that the values associated with these indicators show progress, and thus, the adoption rates may be higher than those reported in this study. Indeed, the Portuguese governmental organization *SPMS – Serviços Partilhados do Ministério da Saúde, EPE*, develops and manages several eHealth projects, which have certainly contributed to this progress over the last two years. To reinforce this idea, the Strategic Plan 2017–2019 [21] expresses the SPMS contribution in order to improve the management of hospitals, circulation of clinical information and articulation with other levels of care and other agents of the sector, betting on the Electronic Health Record (EHR) as an indispensable tool for efficient access management, equity and quality by creating effective conditions for sharing results from complementary diagnostic and therapeutic resources, harmonizing data sets, enhancing clinical research and secondary use of data through interoperability initiatives, as well as information security, cyber security and data privacy.

One of the projects in security context, was launched by the SPMS in 2016 under the name of "Program for Risk Improvement and Information Security", whose objective is to promote the coordination and sharing of good practices related to the information systems of the Ministry of Health. In 2017, the competencies of SPMS, in the context of cybersecurity were reinforced with *Dispatch no. 1348/2017, in Diário da República no. 28/2017, Series II of 2017-02-08*, which identifies a set of new competencies in the coordination, monitoring the implementation and operationalization of good practices for continuous improvement of response to cyber-risks in the health sector.

Regarding the adoption of eHealth Infrastructures, a study of INE (Statistics Portugal) published in Dec 2014 [22], reports that more than 90% of hospitals use the Internet for access to databases and for consultation of procurement catalogues, while 80% use the internet as a means of communication with other hospitals, 72% in internal communication between services and 64% in the training of human resources. In 2014, Internet access in hospitals is universal, 97% with broadband access. 45% of hospitals provide access to the Internet (hotspots) for users, while 35% provide computers with internet access to inpatients. The vast majority of hospitals, 93%, report being present on the Internet, mainly through their own website (87% of those with presence on the Internet) and/or their presence on the website of the Ministry of Health (19%). Of the hospitals present on the Internet, almost all of them (97%) include institutional information on the website, 96% provide an electronic address, and 88% provide information on the services provided.

5 Conclusion

All countries face challenges in modernizing and sustaining high quality health services, and many see effective use of health Infrastructures (as well as best security practices) as central to healthcare transformation [23]. Indeed, there is a tremendous need to learn from and reuse the experiences of others. Doing so requires understanding our position in relation to others. It requires understanding what we have done well that puts us ahead of others and what we should do to reach the level of others. In other words, the understanding of which countries' experiences may be most instructive and helpful. Taking into account the multiplicity of health systems, cultures, and economic capacity, multinational benchmarking in the health sector is always challenging, but the analysis of its results allows us to have a deeper picture of each country's status and progress, facilitating cross-national learning [23]. In this context, the analysis reported in this paper reveals that Portuguese hospitals are above the European Hospitals average, for the adoption of the eHealth Infrastructures. However, their rates are somewhat lower in terms of health safety practices. It has been observed that in recent years there has been an effort by the Portuguese government to promote innovative projects in the health area that suggest a positive evolution of the indicators considered in this study, improving the levels of eHealth Infrastructures and Security.

Finally, the study reported in this paper is not free of limitations. First, the data collected in the study were published in 2014 and their analysis may be slightly outdated. Second, there is a need for further studies to monitor and optimize eHealth Infrastructures and Security in Hospitals for the healthcare system in Portugal. Finally, the scope and size of this paper does not allow to evaluate the level of maturity of the information systems [24, 25] of Portuguese hospitals.

Acknowledgement. We acknowledge the support and sponsorship provided by Porto Polytechnic through the Centre for Research CEOS.PP.

References

1. Carvalho, J.V., Rocha, Á., Abreu, A.: HISMM - hospital information system maturity model: a synthesis. In: International Conference on Software Process Improvement, vol. 537, pp. 189–200 (2016)
2. Fitterer, R., Rohner, P.: Towards assessing the networkability of health care providers: a maturity model approach. Inf. Syst. E-Bus. Manage. **8**, 309–333 (2010)
3. Ludwick, D., Doucette, J.: Adopting electronic medical records in primary care: lessons learned from health information systems implementation experience in seven countries. Int. J. Med. Inform. **78**(1), 22–31 (2009)
4. Jha, A.K., et al.: Use of electronic health records in US hospitals. New Engl. J. Med. **360** (16), 1628–1638 (2009)
5. Ahtonen, A.: Healthy and active ageing: turning the 'silver' economy into gold. European Policy Centre, Europe's Political Economy - Coalition for Health, Ethics and Society (CHES) (2012)
6. Buntin, M.B., et al.: The benefits of health information technology: a review of the recent literature shows predominantly positive results. Health Aff. **30**(3), 464–471 (2011)
7. Mettler, T.: Transformation of the hospital supply chain: how to measure the maturity of supplier relationship management systems in hospitals? Int. J. Healthc. Inf. Syst. Inform. **6** (2), 1–13 (2011)
8. Freixo, J., Rocha, Á.: Arquitetura de Informação de Suporte à Gestão da Qualidade em Unidades Hospitalares. RISTI - Revista Ibérica de Sistemas e Tecnologias de Informação (14), 1–18 (2014)
9. Sharma, B.: Electronic healthcare maturity model (eHMM): a white paper. Quintegra Solutions Limited (2008)
10. NHS, Information Management and Technology (IM&T) Strategy 2012–2017: to provide the knowledge, skills, technology and tools that enable information to be collected, managed, used and shared to deliver excellence in healthcare. Mid-Cheshire Hospitals NHS Found. Trust (2011)
11. Committees of the European Parliament: Communication from the Commission to the European Parliament, the Council, the European Economic and Social Committee and the Committee of the Regions, eHealth Action Plan 2012–2020 - Innovative healthcare for the 21st century. COM 736 (2012)
12. European Commission: European Hospital Survey: Benchmarking Deployment of eHealth Services (2012–2013) Final report. JRC Scientific and Policy Reports (2014)
13. Mikalef, P., Batenburg, R.: Determinants of IT adoption in hospitals - IT maturity surveyed in an European context. In: Proceedings of the International Conference on Health Informatics, Italy (2011)
14. Saleh, M.F.: Information security maturity model. Int. J. Comput. Sci. Secur. (IJCSS) **5**(3), 316–337 (2011)
15. Currie, W., Seddon, J.: A cross-national analysis of eHealth in the European union: some policy and research directions. Inf. Manag. **51**(6), 783–797 (2014)
16. Piha, T.: How Can the EU Drive Ehealth Development? Mede-Tel, European Commission, DG Health and Consumers, Luxembourg (2014)
17. European-Commission: Directive 2011/24/Eu of the European Parliament and of the Council of 9 March 2011 on the Application of Patients' Rights in Cross-Border Healthcare. European Commission, Brussels (2011)
18. epSOS: Smart Open Services for European Patients - WP 1.1: Analysis and Comparison of National Plans/Solutions. European Commission (2009)

19. European-Commission, SIBIS report no. 9: Benchmarking e-Health in the IS in Europe and the US. European Commission (2003)
20. Empirica, Benchmarking ICT Use among General Practitioners in Europe. European Commission: Information Society and Media Directorate General, Bonn (2008)
21. SPMS: Plano Estratégico SPMS 2017–2019 (2016)
22. INE, Inquérito à Utilização das Tecnologias de Informação e da Comunicação nos Hospitais. Destaque: Informação à comunidade social - Instituto Nacional de Estatística (2014)
23. Zelmer, J., et al.: International health IT benchmarking: learning from cross-country comparisons. J. Am. Med. Inform. Assoc. 24(2), 371–379 (2017)
24. Carvalho, J.V., Rocha, A., Vasconcelos, J.B.: Towards an encompassing maturity model for the management of hospital information systems. J. Med. Syst. 39(9), 1–9 (2015)
25. Carvalho, J.V., Rocha, Á., Abreu, A.: Maturity models of healthcare information systems and technologies: a literature review. J. Med. Syst. 40(6), 1–10 (2016)

A Neural Network Based Expert System
for the Diagnosis of Diabetes Mellitus

Oluwatosin Mayowa Alade[1], Olaperi Yeside Sowunmi[1],
Sanjay Misra[1(✉)] (iD), Rytis Maskeliūnas[2], and Robertas Damaševičius[2]

[1] Covenant University, Ota, Nigeria
aladetosinalade@gmail.com, {olaperi.sowunmi,
sanjay.misra}@covenantuniversity.edu.ng
[2] Kaunas University of Technology, Kaunas, Lithuania
{Rytis.maskeliunas,robertas.damasevicius}@ktu.lt

Abstract. Diabetes is a disease in which the blood glucose, or blood sugar levels in the body are too high. The damage caused by diabetes can be very severe and even more pronounced in pregnant women due to the tendency of transmitting the hereditary disease to the next generation. Expert systems are now used in medical diagnosis of diseases in patients so as to detect the ailment and help in providing a solution to it. This research developed and trained a neural network model for the diagnosis of diabetes mellitus in pregnant women. The model is a four-layer feed forward network, trained using back-propagation and Bayesian Regulation algorithm. The input layer has 8 neurons, two hidden layers have 10 neurons each, and the output layer has one neuron which is the diagnosis result. The developed model was also incorporated into a web-based application to facilitate its use. Validation by regression shows that the trained network is over 92% accurate.

Keywords: Expert system · Diabetes diagnosis · Neural network
Back propagation algorithm

1 Introduction

Diabetes is a disease that occurs as a result of the glucose levels being too high due to the absence of, or inadequate amount of insulin in the body. It is a defect in the body's ability to convert glucose (sugar) to energy. Glucose is the main source of fuel for our body. Diabetes develops when the pancreas fails to produce sufficient quantities of insulin (Type 1 diabetes) or the insulin produced is defective and cannot move glucose into the cells (Type 2 diabetes) [1]. Either insulin is not produced in sufficient quantities or the insulin produced is defective and cannot move the glucose into the cells.

The damage that is caused by diabetes can be very severe. It can cause damage to the eyes (blindness), the kidney (kidney failure), the heart (heart attacks), stroke and even lead to amputation [2]. The IDF (International Diabetes Foundation) has stated that 415 million people have diabetes in the world and more than 14 million people have it in Africa. Out of this 14 million Africans, a whopping 1.56 million cases are from Nigeria alone where 1.9% of the adult population has been diagnosed of the same.

© Springer International Publishing AG, part of Springer Nature 2018
T. Antipova and Á. Rocha (Eds.): MosITS 2017, AISC 724, pp. 14–22, 2018.
https://doi.org/10.1007/978-3-319-74980-8_2

Early detection however, can help to bring it under control. The importance of early detection cannot be over emphasized in the treatment of diabetes. It is an important step in the process of recovery for the patient [1]. An early detection of the diabetes mellitus will help in ensuring that the symptoms do not intensify to the level of the severe cases.

The healthcare sector has seen the rise of some of the most promising and groundbreaking startups in the technology world. These innovations have primarily been driven by the advent of software and mobility of digital devices allowing the health sector to automate many of the pen and paper-based operations and processes that currently slow down service delivery. More recently, we're seeing software become far more intelligent and independent. These new capabilities (studied under the banner of artificial intelligence and machine learning) are accelerating the pace of innovation in healthcare. Thus far, the applications of artificial intelligence in health-care have enabled the industry to take on some of its biggest challenges in these areas which include pharmacy, genetics and diagnosis of diseases. Diabetes diagnosis in particular is a complicated process that involves a variety of factors, from the texture of a patient's skin to the amount of sugar that he or she consumes in a day. For as far back as 2,000 years, medicine has used the method of detecting symptoms, where a patient's disease is diagnosed in view of the symptoms that can be spotted (For example, if you have a fever and running nose, you probably have the flu). Be that as it may, the timing of the display of detectable symptoms is too late, especially for deadly ailments like diabetes, cancer and Alzheimer's disease. With artificial intelligence, there is hope that the diseases can be detected well in advance, which will increase the probability of survival (sometimes by up to 90%) [3].

This research is therefore aimed at developing such a system for the diagnosis of diabetes mellitus by leveraging on a neural network based expert system. The remainder of this work is arranged as follows: Sect. 2 discusses the literature review, Sect. 3 explains the methodology while Sect. 4 shows the implementation. The final section concludes the work.

2   Literature Review

The greatest health threats in developed countries are heart disease, cancer, and diabetes [4]. Diabetes particularly is a prevalent disease that has been rising at an increasing rate in a lot of third world countries, including Nigeria. In extreme cases, it is a major cause of severe diseases such as blindness, kidney failure, heart attacks, stroke. In 2012 about 1.5 million deaths were recorded because of diabetes mellitus and another 2.2 million deaths were caused by high blood glucose. WHO projects that diabetes will be the 7th leading cause of death in 2030.

The burden of diabetes on pregnant women is even unique because it can affect both the mother and the unborn child and cause complications in child birth. Diabetes in pregnancy may be neglected due to poor awareness, finances or substandard medical care. This subsequently increases the risk of their offspring getting diabetes, particu-larly if the diabetes in pregnancy is uncontrolled and continues a disturbing trend of diabetes from generation to generation [5]. There is also the issue of lack of adequate medical personnel and facilities specially to compile the necessary tests required to diagnose the disease. All these make for a compelling case as to why a system that can aid the pre-diagnosis of diabetes is needed. The field of artificial intelligence has come to the rescue in this regard, as several expert systems have been built to salvage similar situations.

An expert system can be defined as a program designed to solve problems at a level comparable to that of a human expert in a given domain [6]. It can also be defined as a software system that is able to solve specific complex problems at the level of human experts by the manipulation of accumulated information from the experts following rules that have been set down by the knowledge engineer. The aim of an expert system is to resolve specific issues that would otherwise require these experts through the intelligence gained from them [7]. Expert systems exist in numerous domains, from medical diagnosis to investment analysis and from counselling to production control. Most expert systems have essential components in common [7], which are:

1. The knowledge-base: which stores the knowledge and expertise gathered from experts in that area; information that makes them experts in that category.
2. The Inference Engine: carries out the reasoning through which the expert system reaches a solution.
3. A User Interface: which is the medium through which the expert system can interact with the user. This is done through various methods like dialogue boxes, command prompts etc.
4. Explanation Facilities: This is required to give an explanation as to how and why a particular decision or solution was derived.
5. Working Memory: The working memory for an expert system consists of both facts imputed from questions asked by the expert system to the user, and facts that are deduced by the system. Precise information on a particular problem is represented as case facts and imputed in the expert system's "working memory".

As medical information systems in modern hospitals and medical institutions become larger, it causes great difficulties in extracting useful information for decision support. The manual method of analyzing data has become inefficient, making newer and better methods of computer based analysis essential. It has been proven that the benefits of introducing artificial intelligence into medical analysis are to increase diagnostic accuracy, to reduce costs and to reduce human resources [7]. Artificial Neural Networks (ANN) is currently the next promising area of interest. Already it could successfully apply to various areas of medicine such as diagnostic systems, bio-chemical analysis, image analysis and drug development.

It is one of the techniques used to implement expert systems; a supervisory learning algorithm that models based on the human brain and nervous system. It is composed of artificial neurons and interconnections [8, 14]. Every interconnection (link) has a weight associated with it which determines the strength and the sign of the connection [9].

A neural network consists of layers of similar neurons. It consists of at least an input layer, an output layer and one or more hidden layers. There are several network architectures, with the Feed-forward being the most common because of its versatility. The term feed-forward describes how this neural network processes and recalls patterns. In a feed-forward neural network, each layer of the neural network contains connections to the next layer, the connections extend forward from the input layer to the hidden layer, but no connections move backwards.

The feed-forward neural network can be trained with a variety of techniques from the broad category of back-propagation algorithms, a form of supervised training to optimization algorithms.

2.1   Related Works

[10] made a system that used data mining techniques such as Bayesian Regularization Algorithm, J48 and Radial Basis Function Artificial Neural Networks for diagnosing diabetes type 2. They took advantage of a data set with 768 data samples, 230 of them selected for test phase. Bayesian Regularization Algorithm with 76.95% accuracy outperformed J48 and RBF with 76.52% and 74.34% accuracies, respectively.

Another system that made use of the back-propagation multi-layer artificial neural networks for identifying diabetes mellitus type 2 was done by [11]. Back-propagation is a supervised learning algorithm that works by correcting errors as it "trains" itself to get better. It models the computed output value with the real or expected value and tries to modify the weights according to the calculated error such that the error value keeps reducing till it reaches as little as possible. Sigmoid function was used to train back-propagation and the Pima-Indian database was used. The data set contains 768 data samples, 568 of which were used for training and 268 as testing set. The network developed contained 8 input layer neurons, 6 hidden layer neurons and 2 output layer neurons. Note that the input layer neurons are the eight features which were used in the data set. After 2,000 rounds it reached 82% accuracy.

[12] also made use of the multilayer perceptron (MLP) artificial neural networks for identifying diabetes type 2. In this paper, MLP model includes one input layer with feature of Pima Indians Diabetes, hidden layer with certain neurons and an output layer which has the responsibility of diagnosis. About 20% of data are used as training set, 60% as testing set and finally 20% are used as application set. Time and number of neurons in hidden layer of MLP model are two important parameters. Finally, highest diagnosis accuracy in training phase using MLP model with maximum time and minimum number of neurons in hidden layer in comparing with same times and neuron numbers was 97.61%.

Another set of authors used a dataset with 250 data samples for diagnosing diabetes disease [6]. Each of these 250 data samples consist of 27 features. These features include blood pressure, creatine, pH urine, and fasting blood sugar. Also, the average age of patients in their dataset is between 25 and 78 years. Multi-layer feed-forward artificial neural networks with back- propagation are used for diagnosis. Three training functions namely BFGS Quasi-Newton, Bayesian Regulation and Levenberg-Marquardt are applied in back-propagation algorithm. Finally, back-propagation with Bayesian Regulation function achieved 88.8% of diagnosing accuracy which performed better than BFGS Quasi-Newton and Levenberg-Marquardt functions. Furthermore, data mining techniques with Pima Indians Diabetes dataset is used for identifying diabetes. The applied data mining techniques include SVM, KNN, C4.5 and artificial neural networks with input, hidden and output layers. Finally, artificial neural networks have a higher diagnosing accuracy compared to other data mining techniques [6].

In yet another case, authors use general regression neural networks and Pima Indians Diabetes for identifying type-2 diabetes [13]. A general regression neural network model in this paper is assumed to be a four-layer model; -ne input layer with 8 features from Pima Indians Diabetes, two layers which have 32 and 16 neurons, respectively. Finally, output layer has one neuron. This neuron determines if a person is positive or not. It is used for classification of Pima Indians Diabetes dataset into healthy and patient classes. The above-mentioned dataset with 576 data sample as training set and 192 data set as testing set is used for training and testing processes. The accuracy rate achieved for training and testing phases are 82.99% and 80.21%, respectively. Training phase for diagnosing diabetes type 2 obtained a higher value of accuracy compared to other works studied in this paper [13].

The aim of work however, is to develop a functional neural network based expert system to diagnose diabetes mellitus in women.

3   Methodology

A 4-layer Artificial Neural Network was designed and trained using the Back-propagation method and the Bayesian Regulation (BR) Algorithm. The back propagation learning algorithm works by correcting errors as it "trains" itself to have less errors. It models the computed output value with the expected value (1 or 0) since it has binary output, 1 for true and 0 for false. The BR algorithm was used additionally to avoid over fitting the data set. The dataset was gathered from a medical database called the Pima Indian Database set. It contains 768 rows of data. The individual used were tested on Blood Pressure, Triceps Skin Thickness, Insulin, Body Mass Index, Diabetes Pedigree Function, and age. 500 people tested positive and 268 tested negative to make a total of 768 people.

The Matlab software application was used for the training. Training is the process of modifying the network using a learning mode in which an input is presented to the network along with the desired output. The weights are then adjusted so that the network attempts to produce the desired output. The data set was divided for training, testing and validation. 70% for training, 15% for testing and 15% for validation. The data is trained until it can form a single and accurate output as displayed in the regression graphs in Fig. 2.

3.1   Neural Network Design

The neural network model consists of four layers of neurons as shown in Fig. 1. The input layer with eight (8) neurons, the two hidden layers with ten (10) neurons each, and the output layer with a single neuron.

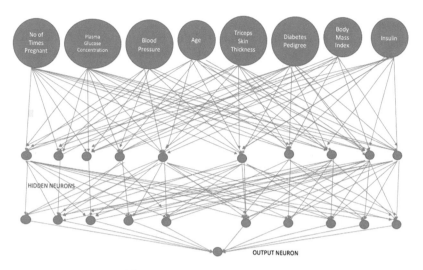

Fig. 1. The neural network model

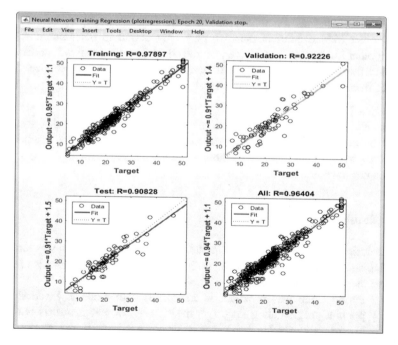

Fig. 2. The regression graph

The Input Layer

This layer consists of eight input neurons which represent the eight inputs that are needed by the system to initiate the diagnosis. These eight inputs represent:

1. The number of times the user has been pregnant;
2. The user's plasma glucose concentration: This is a fasting test which involves abstinence from food and water for a period of 8–10 h before a blood sample is taken. Normal quantity is anything below (6.1 mmol/l) or (101 mg/dl) while concerning quantity is between (6.1 and 6.9 mmol/l) or (111 mg/dl and 125 mg/dl) and diabetic is above (7.0 mmol/l) or (126 mg/dl);
3. The blood pressure of the user (mm Hg);
4. The triceps skin thickness which is at an average of 12 mm for women and 23 mm for the men;
5. Insulin (mu U/ml);
6. Body mass index (weight in kg/(height in m) ^2);
7. Diabetes pedigree function;
8. Age.

The Hidden Layer

Two hidden layers were used with ten neurons each. The hidden neurons have two important characteristics. First, they only receive input from other neurons, such as input or other hidden neurons. Second, they only output to other neurons, such as output or other hidden neurons. Hidden neurons help the neural network understand the input, and they form the output. However, they are not directly connected to the incoming data or to the eventual output.

The Output Layer

There is only one output layer with one neuron which represents the diagnosis result; The neuron that contains the result of the user in decimal form. If the result is 0.5 or greater then, the user has diabetes mellitus while if the result is lower than 0.5 then the user does not have diabetes.

4   Implementation

After the design of the neural network model, a web-based application was built to use the model, so that patients can easily enter their details and get a diagnosis of whether they have the potential to be diagnosed of diabetes or not.

The Training Code was rewritten in JavaScript with Node.js. JavaScript is an incredibly popular programming language, mostly seen in web browsers but gaining popularity in other contexts. On web pages it adds interactivity, from simple animation effects to form validation to full-blown single-page applications.

The interfaces which are the various GUI which the users interact with and also the program modules -operations that can be carried out were also programmed.

The screen shots of the developed system are shown in Figs. 3 and 4. Figure 3 is the home page which is displayed to authorized users of the system after they have logged in. From the home page a user can move on the diagnosis page as shown in

Fig. 4. In this page, the user enters required details needed by the input layer for accurate diagnosis.

Fig. 3. Homepage of the diagnosis system

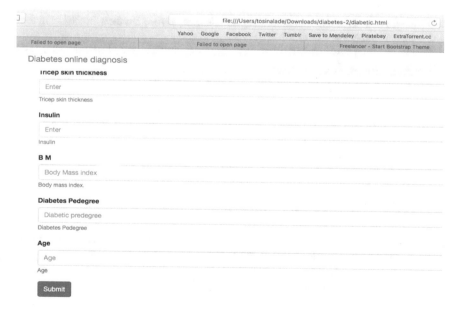

Fig. 4. Diagnosis page

5 Conclusion

This research has developed a neural network model to help in the diagnosis of pregnant women whether they have diabetes or not. The developed model has also been incorporated into a web based application for easy accessibility. As such, health workers or patients at remote locations can access the system, enter patient data in real time and communicate the diagnosis without having to be around the patient.

Some recommendations for its advancement include a Mobile Platform. The system is on a web-based platform at the moment but it would be more beneficial if the system can be based on mobile platforms where there can be offline access on phones, and other devices.

Acknowledgement. We acknowledge the support and sponsorship provided by Covenant University through the Centre for Research, Innovation and Discovery (CUCRID).

References

1. ASPE: Diabetes: A National Plan for Action. The Importance of Early Diabetes Detection. https://aspe.hhs.gov/report/diabetes-national-plan-action/importance-early-diabetes-detection
2. U.S. National Library of Medicine: Medlineplus - Health Topic - Diabetes. https://medlineplus.gov/diabetes.html
3. Gupta, M., Mohammad, Q.: Advances in AI and ML are reshaping healthcare. https://techcrunch.com/2017/03/16/advances-in-ai-and-ml-are-reshaping-healthcare/
4. CIA: CIA. Long-term Global Demographic Trends: Reshaping the Geopolitical Landscape, vol. 5 (2011)
5. Thomas, N.: A Practical Guide to Diabetes Mellitus. Jaypee Brothers Medical Publishers Limited, Tamil Nadu (2016)
6. Kumar, S., Kumaravel, A.: Diabetes diagnosis using artificial neural network. Int. J. Eng. Sci. Res. Technol. **2**(6), 1642–1644 (2013)
7. Negnevitsky, M.: Artificial Intelligence: A Guide to Intelligent Systems. Pearson Education, Harlow (2005)
8. Munakata, T.: Fundamentals of the New Artificial Intelligence. Springer, London (2008)
9. Russell, S., Norvig, P.: Artificial Intelligence. A Modern Approach, 3rd edn. Prentice Hall, Upper Saddle River (2010)
10. Sa'di, S., Maleki, A., Hashemi, R., Panbechi, Z., Chalabi, K.: Comparison of data mining algorithms in the diagnosis of type II diabetes. Int. J. Comput. Sci. Appl. **5**, 1–12 (2015)
11. Olaniyi, E., Khashman, A.: Onset diabetes diagnosis using artificial neural network. Int. J. Sci. Eng. Res. **5**, 754 (2014)
12. Al-Mufad, N., Al-Hagery, M.A.: Using prediction methods in data mining for diabetes diagnosis, vol. 1 (2014)
13. Kayaer, K., Yildirim, T.: Medical diagnosis on Pima Indian diabetes using general regression neural networks, vol. 6 (2003)
14. Lino, A., Rocha, Á., Sizo, A.: Virtual teaching and learning environments: automatic evaluation with artificial neural networks. Clust. Comput., 1–11 (2017)

Developing a Calorie Counter Fitness App for Smartphones

Adewole Adewumi[1,2], Godwin Olatunde[1], Sanjay Misra[2(✉)] [ID],
Rytis Maskeliūnas[3], and Robertas Damaševičius[3]

[1] Department of Computer and Information Sciences,
Covenant University, Ota, Nigeria
wole.adewumi@covenantuniversity.edu.ng,
olatundegodwin@gmail.com
[2] Center of ICT/ICE Research, CUCRID Building,
Covenant University, Ota, Nigeria
sanjay.misra@covenantuniversity.edu.ng
[3] Kaunas University of Technology, Kaunas, Lithuania
{Rytis.maskeliunas,robertas.damasevicius}@ktu.lt

Abstract. A number of mobile fitness devices as well as smart watches have emerged on the technology landscape. However, the rate of adoption of these devices is still low especially in developing countries with a teeming population. On the other hand, smart phones are becoming ubiquitous given their steady price decline. To this end, the present study aims to leverage the smartphone platform by developing a smart phone fitness app that tracks the calories burnt by individuals who go about their daily activities while carrying their smart phones with them. In order to achieve this, the design specification for the application was done using Unified Modeling Language diagrams such as use case diagrams and sequence diagrams. This was then implemented using the following tools: Angular - a JavaScript framework and Ionic - a hybrid framework that was hosted via the Heroku Cloud Application Platform. The initial results show that the app can gain traction in terms of its adoption given the fact that it is cheaper to download the app than buy a new smart watch for the same purpose.

Keywords: Android · Calorie counter · Fitness app · Smart phones

1 Introduction

Health tracking systems can be categorized based on technological advancement into two groups namely: traditional and electronic health tracking systems [1]. In times past, individual's health activity was measured in an analogue manner requiring the use of analogue scales and thermometers [2]. As the readings were taken, they were recorded in the file of the individual at a given medical centre [3]. Such records were only found in the health facility where the readings were taken. The drawbacks of this approach includes among others [4, 5]: the difficult to comprehend handwriting of a number of health practitioners, the records soon become bulky and duplications are often

© Springer International Publishing AG, part of Springer Nature 2018
T. Antipova and Á. Rocha (Eds.): MosITS 2017, AISC 724, pp. 23–33, 2018.
https://doi.org/10.1007/978-3-319-74980-8_3

inevitable. In addition, the records are not always readily available and accessible unless a visit is made to the medical centre.

Electronic health tracking systems overcame the limitations of the traditional health tracking systems and also heralded the advent of fitness tracking devices [6]. As of today, the use of stopwatches and bathroom scales to track fitness level is gradually becoming a thing of the past in both developed and developing nations of the world [7]. A fitness tracker or action tracker is a gadget or application for observing and following fitness-related measurements, for example, distance covered in walks or run, calorie utilization, and at times pulse and nature of rest [8]. The term is basically used to describe devices that are connected, a significant part of the time remotely, to a PC or PDA for whole deal data taking after, an instance of wearable technology [9, 10, 11]. The expression "activity trackers" now essentially alludes to wearable gadgets that screen and record an individual's fitness action [12]. The idea came into being out of composed logs that prompted spread sheet-style PC signs in which passages were made physically. Improvements in innovation in the late 20th and mid-21st century permit robotizing the observation and recording of fitness exercises and incorporating them into all the more effectively worn gear [13]. Early cases incorporate wristwatch-sized bike PCs that checked attributes like speed and length. Wearable heart rate screens for competitors were accessible in 1981 [14]. Wearable fitness GPS beacons, including remote heart rate checking that coordinated with business review fitness gear found in exercise centres, were accessible in shopper review hardware by the mid-2000s [15].

Health-tracking platforms use sensors in a user's activity tracker or mobile device to record physical fitness activities for example walking or cycling, which are then tallied up against the user's health goals to provide a comprehensive view of their fitness [16]. Users can choose whom to share their health tracking data with as well as delete the data at any point in time. In this regard, fitness tracking devices and wearables - especially watches are now trending. However, the rate of adoption of these devices is low especially in developing countries with a teeming population such as Nigeria (in sub-Saharan Africa) [17]. One reason for this low adoption among others is the high cost of the wearables [18]. On the other hand, mobile phones are quite ubiquitous as a result of the steady decline in price and the steady rise and improvement in mobile software and apps [19, 32]. This study therefore explores the possibility of developing a mobile application that would serve as a substitute to the use of these wearables in tracking calorie burn in individuals.

The rest of this study is structured as follows: Sect. 2 discusses related works while Sect. 3 presents the design specification of the proposed application. In Sect. 4, the design specification is implemented as a mobile application. Section 5 discusses the results obtained while Sect. 6 concludes the paper.

2   Related Works

The study in [20, 21], suggested that young, currently healthy adults, have some interest in apps that attempt to support health-related behaviour change. They also valued the ability to record and track behaviour and goals and the ability to acquire advice and information "on the go". While the study in [22, 23] demonstrated that apps promoting physical activity applied an average of 5 out of 23 possible behaviour change techniques.

The study in [24] observed that paid apps tended to be user-friendlier in terms of the user's literacy in comparison to free apps hence the need to explore the development of more user-friendly and accessible apps. This lead to the study in [25] which showed evidence that apps are a feasible and acceptable means of administering health interventions. As a result, the study in [26] opined that the use of particular design features and application of evidence-based behaviour change techniques could optimise continued use and the effectiveness of internet/smart phone interventions. Furthermore, the study in [27] revealed that abandonment does not necessarily reflect individuals' dissatisfaction with technology. Individuals were selling their old devices because they achieved their goals or were upgrading to newer models, scenarios that indicate success, rather than failure of technologies.

In addition, the study in [28] found that there was an association between use of calorie and fitness trackers and eating disorder symptomatology. This view was supported in [29] where it was revealed that mobile health app users had significantly higher scores for eating behaviour than nonusers, and the impact of using more than one type of mobile health app significantly improved eating behaviour. Most participants also identified app use with feeling healthier, better self-monitoring of food intake and exercise, and having more motivation to eat healthier and increase physical activity.

As a result, the study in [30] compared four popular fitness trackers through a combination of subjective and objective experimental results as well as usability evaluation from seven real users and the results showed that Withings Pulse (which costs $120) was the most acceptable in terms of price and satisfaction levels. Another study in [31] developed a mobile app that provides advice about obtaining a healthy diet according to age, clinical history and physical condition. The app was validated through usability evaluation and was found to be easy to use and attractive.

3 Design of the Proposed App

Various materials such as journals, articles, conference proceedings, books and research papers were reviewed in order to analyse and identify the existing fitness tracking apps and to also gather requirements for the proposed mobile application. Unified Modelling Language diagrams were drawn to conceptualize the application that was to be built. These include the use case diagram, sequence diagram and state chart diagram. Figure 1 depicts the use diagram of the proposed application comprising of five use cases namely: track calorie burnt, calculate steps taken, measure distance covered, input/edit weight and input/edit height.

From Fig. 1, it can be observed that the user (the actor) has access to all the five use cases. The sequence of activities especially as regarding the order of occurrence is depicted in Fig. 2.

From Fig. 2, the logged in user launches or starts the application and can get an overview of what the application is about as a first time user. All the records of new users are stored in the database for future retrieval. The calorie burn rate is determined

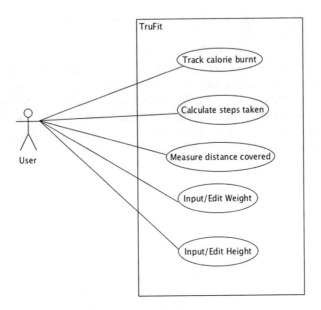

Fig. 1. Use case diagram for the proposed system

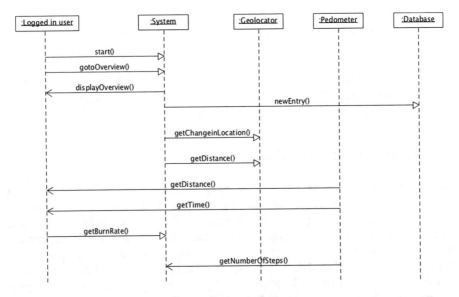

Fig. 2. Sequence diagram

using three elements, which include: the step count, the distance covered and the time taken. Figure 3 depicts the state diagram of the proposed application. To access the system, the user in question would have to be registered in the application's database by filling in some biodata.

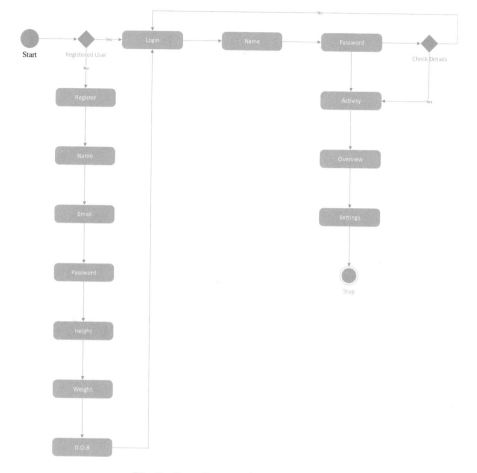

Fig. 3. State diagram of the proposed app

The mobile application was developed using PHP programming language. HyperText Markup Language (HTML) was used for the web mark-up (structure) and Cascading Style Sheets (CSS) was used for styling. Animations were added to the app to enhance interactivity using AngularJS – a JavaScript framework. Apache HTTP Server was used to handle the communication between the servers and the database.

4   Implementation

This section shows the outcome of the design specification conducted in the previous section. Figure 4 shows the page for authorized persons to login in so as to access their dashboard.

Fig. 4. Login page

From Fig. 4, the user will not be able to proceed without first registering as a user. The registration process does not cost anything. Figure 5 shows the page for unauthorized persons to create an account in order to access their dashboard.

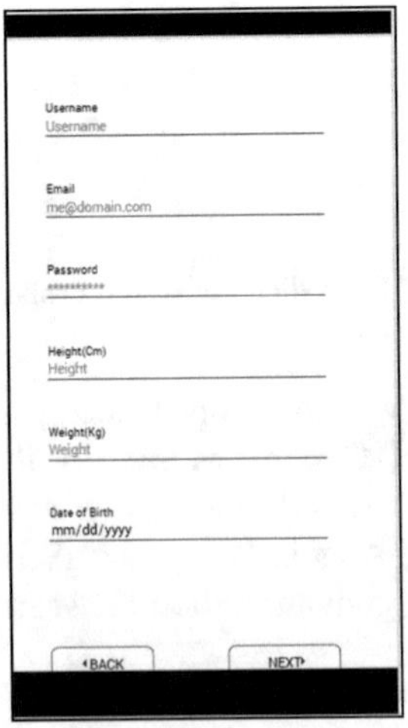

Fig. 5. Register page

Once a user is registered and successfully logs in to the application, s/he can then access and see the history/record in terms of total number of steps s/he has taken, the distance covered and the total calories burnt as depicted in Fig. 6.

Fig. 6. Overview page

From Fig. 6, we see that the logged in user has made 2,456 steps and burnt 42 kcal. The total distance covered by the user is given as 7,256 km. Figure 7 shows the stopwatch to start an activity. When the start button is tapped, it starts recording step counts and calories burnt per time.

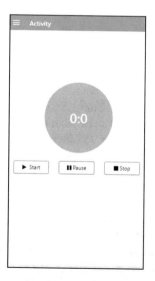

Fig. 7. Activity page

Fig. 8. Side menu page

The other feature of the application is the side menu, which contains the various options and pages available in the mobile application as depicted in Fig. 8.

From Fig. 8, we observe that the content of the side menu page includes: Activity, Overview, Settings and Logout.

5 Discussion

This research project is significant to the field of medicine especially as it relates to the fitness of the human body. Developing the application as an Android-based app improves its accessibility via smartphones given the wide spread adoption of Android-based mobile devices. In addition, there is cost savings for the consumer, as there would be no more need to purchase a wearable. The mobile application developed tracks fitness data over time; calculates the number of calories burnt while exercising; monitors weight loss and number of steps taken; and improves the overall fitness of an individual unlike the application developed in [31] which basically provides advice about obtaining a healthy diet.

Calorie Fit-Burn trackers have seen a lot of advances in the health and fitness sector of technology. This study, thriving on the recommendations of earlier researchers has decided to make it standalone, hence making it easily accessible at once to every average person who cannot afford to purchase a fitness band or an additional hardware, it would help improve good standard of living and in turn reduce health diseases and death brought about by the lack of exercise. This application has been built with some of the latest technologies hence; it enjoys the security benefits that come with those time-tested technologies. When deployed, it would greatly help users achieve their fitness and health goals. With such information as knowing the amount of calories burnt per activity makes for room to calculate the calorie intake and maintain a healthy and balanced life.

6 Conclusion

The mobile Calorie Fit-Burn Tracker has been built using Angular (a framework of JavaScript) compiled with Ionic Framework which allows for compiling across all mobile platforms i.e. Android, iOS, and Windows mobile. It is believed that the application would give rise to healthy living amongst the average of people, due to the plus side that it does not require an additional hardware (bands or watches) to function with. This project has been deployed as a mobile application to allow easy access for users. However, with the rapid growth in the use of cheaper fitness bands, it would be profitable for such platforms to be exploited. This would bring about greater portability and access to a wider range of persons. The following activities are planned for future work:

- Conduct of usability evaluation to obtain feedback from users
- Integration with Smart Watches and Devices: Due to the increase in smart watches and health fitness bands. According to Moore' Law, fitness bands and electronic devices would get cheaper along the years with better performance. Hence the need of an integration with fitness bands and smart watches
- Implementation of various Activities: This current platform only caters for walking and running; a provision needs to be made to accommodate other physical activities.
- Setting Workout Goals: Users should possess the ability to set out their desired workout rate, exercises and weight loss goal in which they aim to achieve.

Acknowledgement. We acknowledge the support and sponsorship provided by Covenant University through the Centre for Research, Innovation and Discovery (CUCRID).

References

1. Lentferink, A.J., Oldenhuis, H.K., de Groot, M., Polstra, L., Velthuijsen, H., van Gemert-Pijnen, J.E.: Key components in eHealth interventions combining self-tracking and persuasive eCoaching to promote a healthier lifestyle: a scoping review. J. Med. Internet Res. **19**(8) (2017)
2. Wiener, L., Battles, H., Zadeh, S., Widemann, B.C.: Validity, specificity, feasibility and acceptability of a brief pediatric distress thermometer in outpatient clinics. Psycho-Oncology **26**(4), 461–468 (2017)
3. King, H.H., Cayce, C.T., Herrin, J.: Thermography examination of abdominal area skin temperatures in individuals with and without focal-onset epilepsy. Explore: J. Sci. Healing **13**(1), 46–52 (2017)

4. Samuel, V., Adewumi, A., Chuma-Ibe, K., Misra, S., Omoregbe, N.: Design and implementation of a clinical pharmacy management and electronic prescription system. In: Advances in Digital Technologies: Proceedings of the 8th International Conference on Applications of Digital Information and Web Technologies ICADIWT 2017, vol. 295, p. 171. IOS Press (2017)

5. Okuboyejo, S., Akor, S., Adewumi, A.O.: Development of a mobile feedback system for health institutions in Nigeria. Afr. J. Comput. ICT **5**(6), 41–44 (2012)

6. Jin, H., Abu-Raya, Y.S., Haick, H.: Advanced materials for health monitoring with skin-based wearable devices. Adv. Healthcare Mater. (2017)

7. Lox, C.L.: The Psychology of Exercise: Integrating Theory and Practice. Taylor & Francis, Routledge (2017)

8. Hulsey, N.L.: For Play: Gamification and Everyday Life. North Carolina State University, Raleigh (2015)

9. Yaseen, M., Saleem, K., Orgun, M.A., Derhab, A., Abbas, H., Al-Muhtadi, J., Iqbal, W., Rashid, I.: Secure sensors data acquisition and communication protection in eHealthcare: review on the state of the art. Telematics Inform. (2017)

10. Soh, P.J., Vandenbosch, G.A., Mercuri, M., Schreurs, D.M.P.: Wearable wireless health monitoring: current developments, challenges, and future trends. IEEE Microw. Mag. **16**(4), 55–70 (2015)

11. Chen, C., Knoll, A., Wichmann, H.E., Horsch, A.: A review of three-layer wireless body sensor network systems in healthcare for continuous monitoring. J. Mod. Internet Things **2**(3), 24–34 (2013)

12. Hendler, J., Mulvehill, A.M.: Social Machines: The Coming Collision of Artificial Intelligence, Social Networking, and Humanity. Apress, New York City (2016)

13. Maqbool, A., Mohsin, N., Siddiqui, H.: Future application trends for health based internet of things. Int. J. Comput. Appl. **118**(18) (2015)

14. Bloss, R.: Wearable sensors bring new benefits to continuous medical monitoring, real time physical activity assessment, baby monitoring and industrial applications. Sens. Rev. **35**(2), 141–145 (2015)

15. Casselman, J., Onopa, N., Khansa, L.: Wearable healthcare: lessons from the past and a peek into the future. Telematics and Inform. (2017)

16. Murray, S.: Get Smarter: The Wearables, Carriables, and Shareables of Digital Self-Actualization. The University of Wisconsin-Madison (2016)

17. Idowu, P.A.: Information and communication technology: a tool for health care delivery in Nigeria. In: Computing in Research and Development in Africa, pp. 59–79. Springer International Publishing (2015)

18. Shih, P.C., Han, K., Poole, E.S., Rosson, M.B., Carroll, J.M.: Use and adoption challenges of wearable activity trackers. In: IConference 2015 Proceedings (2015)

19. Osebor, I., Misra, S., Omoregbe, N., Adewumi, A., Fernandez-Sanz, L.: Experimental simulation-based performance evaluation of an SMS-based emergency geolocation notification system. J. Healthcare Eng. (2017)

20. Dennison, L., Morrison, L., Conway, G., Yardley, L.: Opportunities and challenges for smartphone applications in supporting health behavior change: qualitative study. J. Med. Internet Res. **15**(4) (2013)

21. Gowin, M., Cheney, M., Gwin, S., Franklin Wann, T.: Health and fitness app use in college students: a qualitative study. Am. J. Health Educ. **46**(4), 223–230 (2015)

22. Middelweerd, A., Mollee, J.S., van der Wal, C.N., Brug, J., te Velde, S.J.: Apps to promote physical activity among adults: a review and content analysis. Int. J. Behav. Nutr. Phys. Act. **11**(1), 97 (2014)

23. Chen, J., Cade, J.E., Allman-Farinelli, M.: The most popular smartphone apps for weight loss: a quality assessment. JMIR mHealth uHealth **3**(4) (2015)
24. Charlene, C.A., Graff, K., Harris, J.K., McQueen, A., Smith, M., Fairchild, M., Kreuter, M.W.: Evaluating diabetes mobile applications for health literate designs and functionality. Prev. Chronic Dis. (2015/14_0433) (2014). http://www.cdc.gov/pcd
25. Payne, H.E., Lister, C., West, J.H., Bernhardt, J.M.: Behavioral functionality of mobile apps in health interventions: a systematic review of the literature. JMIR mHealth uHealth **3**(1) (2015)
26. Tang, J., Abraham, C., Stamp, E., Greaves, C.: How can weight-loss app designers' best engage and support users? A qualitative investigation. Br. J. Health. Psychol. **20**(1), 151–171 (2015)
27. Clawson, J., Pater, J.A., Miller, A.D., Mynatt, E.D., Mamykina, L.: No longer wearing: investigating the abandonment of personal health-tracking technologies on craigslist. In: Proceedings of the 2015 ACM International Joint Conference on Pervasive and Ubiquitous Computing, pp. 647–658. ACM (2015)
28. Simpson, C.C., Mazzeo, S.E.: Calorie counting and fitness tracking technology: associations with eating disorder symptomatology. Eat. Behav. **26**, 89–92 (2017)
29. Sarcona, A., Kovacs, L., Wright, J., Williams, C.: Differences in eating behavior, physical activity, and health-related lifestyle choices between users and nonusers of mobile health apps. Am. J. Health Educ. **48**(5), 298–305 (2017)
30. Kaewkannate, K., Kim, S.: A comparison of wearable fitness devices. BMC Public Health **16**(1), 433 (2016)
31. de la Torre Díez, I., Garcia-Zapirain, B., López-Coronado, M., Rodrigues, J.J., del Pozo Vegas, C.: A new mHealth app for monitoring and awareness of healthy eating: development and user evaluation by spanish users. J. Med. Syst. **41**(7), 109 (2017)
32. Vázquez, M.Y.G., Sexto, C.F., Rocha, Á., Aguilera, A.: Mobile phones and psychosocial therapies with vulnerable people: a first state of the art. J. Med. Syst. **40**(6), 157 (2016)

Information Technology in Education

Triple-Loop Learning Based on PDSA Cycle: The Case of Blogs for Students-at-Risk

Ahmad F. Sad[1], Andreea Ionica[2] , Monica Leba[2(✉)] ,
and Simona Riurean[2]

[1] Carmel College, Acre, Israel
ahmadsaad2011@gmail.com
[2] University of Petrosani, Petrosani, Romania
{andreeaionica, monicaleba}@upet.ro,
sriurean@yahoo.com

Abstract. We are witnessing the evolution of the concepts capturing the dynamics of learning. Starting with "following the rules" in the single-loop learning model, the concepts have evolved showing the need for "changing the rules" and for "thinking outside the box" with significant changes also at the "mental model". The idea of continuous improvement is summarized in "learning about learning", the essence of triple-loop learning model "reflexive thought and outside contribution" in order to change and transform organizational practices and learning processes. In this paper, we discuss the particular case of the students at risk and their learning improvement possibilities using ICT tools, namely blogs.

Keywords: ICT · Continuous improvement · Web 2.0 technologies

1 Introduction

Learning takes place on several levels: from single-loop learning (adaptive learning) through double-loop learning (reflection in and on action) to triple-loop learning (meta-learning), and extending ones understanding and competencies of how to learn individually and in groups [1]. The literature refers to consistent approaches to triple-loop learning [2], to the mixed methodologies between systematic approaches behind them or to the integrated approaches based on Checkland, Eden and Forrester. Our research uses Deming's PDSA cycle as a foundation and support to develop a new approach on triple-loop learning customized and validated for students at risk use of blogs.

Schools focus on ensuring that all students succeed in life and participate effectively in society. Yet there are students who are at risk of dropping out due to many factors. The term "At-Risk" is used to describe students who are in danger of not meeting educational goals such as graduating from high school or acquiring the skills necessary to become contributing members of society, some of them exhibit disruptive behavior that interferes with their learning. Their background characteristics may place them at or below the poverty level. Other characteristics include low grades and tests scores, abundant absences from school. At-risk students usually feel that they are

© Springer International Publishing AG, part of Springer Nature 2018
T. Antipova and Á. Rocha (Eds.): MosITS 2017, AISC 724, pp. 37–46, 2018.
https://doi.org/10.1007/978-3-319-74980-8_4

overwhelmed by the content covered in high school. They may have learning disabilities that make reading and writing difficult for them. So how can we motivate these students? How do we know that they became motivated? According to [3] once they are motivated they pay attention, they begin working on tasks immediately, they ask questions and volunteer answers, and they appear to be happy and eager. Technology can be used to motivate these students [4, 5], research reveals how feelings of autonomy, extrinsic and intrinsic goal orientation, and task value are related to increased motivation among at-risk students [4]. Information and Communications Technologies (ICT) have been widely perceived as the lever that would lead to significant educational and pedagogical outcomes and support students' development on the knowledge and skills needed to succeed in the 21st century society, where the graduates of secondary school needed to fulfil the digital literacy requirements (i.e. ICT skills, Critical thinking skills, and ethical skills). However, the results of technology initiatives have been mixed. Often, the introduction of technology into classrooms has failed to meet the grand expectations proponents anticipated [6]. If achievement effects of traditional versus computer-assisted instruction can produce similar positive or negative effects, then inquiries into the types of highly-effective technology and supporting classroom infrastructures are essential to strategic planning, particularly in regards to the literacy instruction of at-risk student populations [7]. Web 2.0 applications including blogs, wikis, social networking, social bookmarking, RSS, podcasting, media sharing etc., have enabled students to master many parts of the digital literacy requirements. Academics, researchers, educators and policymakers have advocated that the emerged Web 2.0 applications have the potential to offer enhanced learning opportunities for both students and educators and support lifelong competence development [8]. Among the Web 2.0 technologies and applications, using blogs in the classroom can help increase student learning using student's preferred learning style, personal interest, and engagement. It also encourages self-reflection for the student and critical thinking. The on-line fast publication of a blog and a whole world audience increases student motivation for writing. Student Blogging bridges that gap between home and the classroom and creates an unlimited learning environment. It allows collaboration that promotes constructive environment. And for low-achieving students blogging can give the "silent student" a voice by allowing them the opportunity to write on topics of interest [9]. Teachers also play a role in motivating their students. Teachers who perceive themselves efficacious will spend more time on student learning, supporting students in their goals and reinforcing intrinsic motivation [10]. Teachers with a high sense of efficacy feel a personal accomplishment, have high expectations, feel responsibility for student learning, have strategies for achieving objectives, a positive attitude about teaching and believe they can influence student learning [11].

The problem to be solved in the particular case of using blogs is to improve the performance of students-at- risk. After validating the approach, the goal of the study is to understand which factors, related to school science, can interfere with engagement of students-at-risk of dropping-out and to know what kind of activities and teaching strategies are adequate to these students. The proposed solution will combine two advantages; first, it promotes motivation by using ICT and blogs from Web 2.0 technologies, and secondly it enhances digital literacy requirement for secondary students through practicing ICT skills, Critical thinking, and self-reflection. All those factors will promote lifelong learning.

2   Conceptual Background and Issues in Students-at-Risk Using Blogs Case

Students motivation: Motivation is a significant factor in students' academic lives; it affects their classroom behavior, and as a result, future success. Motivated behavior is defined by [12] as being, "energized, directed and sustained". Student behavior is influenced by two major kinds of motivation–namely, autonomous and controlled. The most influential factors on students' motivation are family or social factors [13], school and peer interaction factors, and teachers and teaching styles.

It was stated that students perceived their teachers as the most influential factor in motivating them for their learning, and for this reason, students need their teachers to encourage them by applying different motivating techniques like problem-solving and inquiry-based instructions [14]. The problem as defined by professional sources and educational literature led to the development of the following three interventions: student autonomy, goal setting, and positive teacher feedback. By offering a greater amount of choices to the students, and by providing authentic assessments, and allowing students to take a more active role in their education. This will improve intrinsic motivation of secondary school students. It will instill a mentality of learning for mastery as opposed to extrinsic rewards.

Preparing students for the 21st century skill through ICT and Web 2.0 Technologies: In [15] High school students spend much of their educational journeys immersed in "old" literacies of paper, pencil, and print books. But outside of the classroom, they are exposed to information and communication technologies–such as blogs, wikis, Internet browsers, multimedia, social networking sites, and a wide range of software each of which demand new literacies. This disconnect is a serious problem for schools because it reflects a decline in school's relevancy to students' futures, and the gap between how schools teach and how students learn will only grow over time. It may be understandable for teachers and leaders to be a bit behind in the use of technology, but it is no longer adequate or appropriate to hold students back. Positive attitudes towards computers are positively correlated with teachers' levels of experience with computer technology and are recognized as a necessary condition for the effective use of ICT in the classroom [16]. The new required Literacies mentioned that each student must graduate from high school with the "new" basic skills for life in the 21st century, they include: (1) innovation and imagination; (2) communication, collaboration, teamwork, and critical thinking skills; (3) adaptability and agility; (4) interactivity and information analysis; and (5) initiative and self-direction.

So schools need to be prepared to use ICT in their education because ICT is a major component of economic growth in all countries. Given that young people today need to be skilled in using these technologies as students, job-seekers or workers, consumers and responsible citizens, those who have no access to or experience in using ICT will have it increasingly difficult to participate fully in economic, social and civic life. However, basic ICT skills may not add value unless they are well paired with cognitive skills and other skills, such as creativity, communication skills, teamwork and perse-verance. Schools need sufficient ICT resources to help students both to learn how to use and benefit from these technologies and to acquire new knowledge and skills, in other subjects, through using them. ICT can also help teachers and school administrators to work more efficiently. A distribution of resources across and within education systems has long been an important issue for both equity and excellence in education. Given the rapid advances in technology and the central role ICT now plays in all aspects of life, education policy makers need to consider how to ensure that ICT resources and stu-dents' access to those resources are provided equitably within education systems.

ICT and motivation: It is generally accepted that intrinsic student motivation is a critical requirement for effective learning but formal learning in school places a huge reliance on extrinsic motivation to focus the learner. This reliance on extrinsic moti-vation is driven by the pressure on formal schooling to "deliver to the test." The experience of the use of ICT in formal learning is marked with a naive and largely unfulfilled assumption that it would of itself promote a "game-changing" shift in student motivation [17]. Lawlor and his team investigated the effectiveness of a team-based, technology-mediated model called Bridge21 to encourage intrinsic student motivation. It is not enough, therefore, to simply place work on a laptop, teachers must shift their deep-rooted preferences for worksheets, lecture, and assessment to include alternative approaches to learning through meaningful creation, social media, and project driven curriculum. But for successful results in integrating ICT the curriculum should be designed based on the students' learning preferences or prior knowledge. As for students at risk, ICT can bring more to motivate them: as [18] research showed that. Activities were structured to incorporate individual accountability, positive interde-pendence and interaction, collaborative skills, and group processing. Motivation, self-confidence, learning attitudes, and achievement were improved.

Motivation and using a Blog as an E-Portofolio: In [19] the study examined an electronic portfolio design based on blog services and program called blog folio. Results showed that students expressed the feature of easy to use and their willingness to maintain their portfolios. Portfolios using Web 2.0 technology can be maintained much easier and updated much faster. It can include multimedia files like graphs and audio/video clips. It is much easier for teachers to view blog folios to many students and provide feedback. Teacher can subscribe to RSS reader in order to get immediately updates about changes in students portfolios. It was confirmed that collaborative learning has positive impact on students learning, so viewing peers portfolios can be considered one type of collaborative learning where students have a model to reflect upon and learn. Building a portfolio with personal style in simple steps will enhance students' motivation in maintaining their portfolios. As [20] puts it: blogs can be used as supplementary mediums to promote achievement and knowledge acquisition of students as well as information searching and sharing skills within a learning community.

Using e-portfolio system individually positively influenced some subjects such as encouraging students using computer, reaching the information on virtual environ-ments, assessing themselves and monitoring their process to computer skill, and developing [21]. E-portfolios provide a creative way of organizing, summarizing, and sharing student artifacts, and demonstrating evidence of students' professional growth [22]. Evaluators can use these easily accessible portfolios to gauge student performance while the portfolios themselves permit rapid and immediate inclusion and reconfiguration of portfolio data by the owner. Electronic portfolios are also useful for accreditation purposes to measure and assess student learning outcomes.

E-portfolio added value in pupils' deep learning and keep them engaged, it acted as a medium to involve parents, promoted pupils' self-esteem, and was acknowledged as a valuable assessment tool and a challenge for the school community [23]. There are many programs that can be used to create an e-Portfolio, Mahara is an open source e-portfolio web application that allows users to keep a portfolio online by creating, uploading and linking content. Learners can provide each other feedback on their portfolios and collaborate in groups on common projects using forums and creating group portfolios. Another program for creating e-Portfolio is integrated an e-portfolio application on Facebook, which is one of the most popular social networks of the present-day and has been made available the use of for students. It was shown that e-portfolio application is a process in which students were creating original products, developing the skills of using technologies and having low grade anxiety.

3 A New Approach on Triple Loop Learning Model Based on the PDSA Cycle

In Fig. 1 there are presented: First loop (I) means doing things right, or test and track, incremental learning; Second loop (II) means doing the right things, or evaluate and adapt, reframing; Third loop (III) means decide what is right, or learn and share (conceptualize the plan), transformational learning (Table 1).

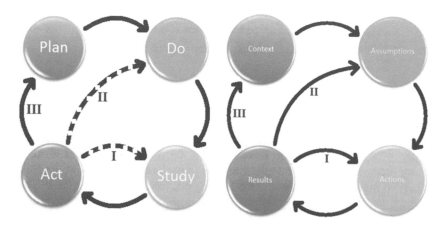

Fig. 1. PDSA versus triple loop learning.

Table 1. Relation between PDSA cycle and triple loop learning.

Plan	Context
Formulating a theory, defining goals and metrics	Formulate learning strategies. Process for generating mental maps and goals for learning in a curriculum context: • Increased learner effectiveness or performance gains – The degree of learning efficiency • More positive student attitudes to learning – The degree of learner engagement and satisfaction
Do	**Assumptions**
Implementing plan components	Standard representations Mental maps and models Set implementation hypotheses for the established goals using ICT & Web 2.0 tools: • Select and apply ICT systems and devices effectively • Utilize common generic software tools in their private lives • Use specialized tools for work • Flexibly adapt to changes in infrastructure and applications
Study	**Actions**
Monitoring to test plan validity for pluses and minuses of students progress and further improvement	Practices and behaviors Implement the ICT tools and technics in learning (e.g. WiKis, Blogs, Social Networking, Social Bookmarking, RSS, podcasting, media sharing)
Act	**Results**
Integrating the learning generated by the entire process and adjusting the theory, methods, goals and purposes	Feedback for the next loop of learning based on consequences and outcomes, from the well-known key domains: Cognitive: critical thinking, problem solving, information literacy; knowledge; creativity Intra-personal: intellectual openness; positive self-evaluation Inter-personal: communication; collaboration (teamwork); leadership.

4   The Case of Students-at-Risk and Blogs

The research intends to examine students over four years including six semesters and their responses to the use of blogs and discussion boards as a key part of learning and reflection. The study adopts the classroom and the organization, looks at how the blogging process moves students from a single-loop learning process to double-loop learning and reflection, and enhances the learning and reflection for the educator, which contributes to the third loop by enhancing the organizational learning approach and demonstrating an increased student satisfaction as measured by student evaluations and increased self-reflection on content specific knowledge, improved individual learning and overall classroom learning.

First loop: The students do their homework online using blogs for "doing things right". They are tested and their posts on blogs are evaluated by the teacher (Fig. 2).

Fig. 2. Blogs of students.

Second loop: The findings showed that students were able to demonstrate their interconnected technological pedagogical knowledge in their reflection to justify their competencies (Fig. 3).

Fig. 3. Class blogs for the last 4 years.

Students have the opportunity to do their homework, to explore more than what they have been taught and asked as homework for "doing the right things". Topics are subject to dialogue between student and teacher. There is ongoing assessment and adaptation of both the student and the teacher.

Third loop: This shows the importance of providing a discipline-specific framework to guide students' reflection of learning. The student and the teacher are partners in the teaching-learning process for "deciding what is right", learn and share. The best evaluation is done through teacher efficacy (Table 2).

Table 2. Prerequisites for third-loop model operation.

Three conditions that enable collective teacher efficacy to flourish	
Advanced teacher influence	Extent of teacher leadership in a school matters Providing teachers greater autonomy and influence over important decisions
Goal consensus	Clear set of goals Consensus between teachers Commitment to achieve these goals
Responsiveness of leadership	Help others carry out their duties effectively Responsive and show concern and respect for their staff Protect teachers from issues and influences that detract from their teaching time or focus Responsiveness requires awareness of situations

5 Conclusions

Our applied model is related to:

- Solving the identified problems (related to students at risk)
- Increasing process performances with elements of student motivation and teacher effectiveness (related both to student and teacher perspective)
- Continuous improvement of learning processes in the context of change of organizational practices (Table 3).

Table 3. Results of triple loop learning.

Recommendation	Description	Solution
Improve school leadership	It is the starting point for the transformation Leaders are selected in an un-appropriate methods They are unprepared to exercise their functions They lack of the support they need to succeed in their work	School leadership preparation programs Coaching, mentoring and networks Provide good working conditions, systemic support and incentives
Enhance school climate for learning	Low performing disadvantaged schools are at risk of difficult environments for learning. Policies specific for these schools	Prioritize the development of positive teacher-student and peer relationships Promote the use of data information systems for school diagnosis Alternative organization of learning time Smaller schools and smaller classes

(continued)

Table 3. (*continued*)

Recommendation	Description	Solution
Enhance quality of teachers	Disadvantaged schools lack quality teachers	Providing targeted teacher education Providing mentoring programs Developing supportive working condition
Provide effective classroom learning strategies	There is evidence that certain pedagogical practices can make a difference for students at risk students	Improve learning in classrooms Use diagnostic tools and formative and summative assessments Ensuring that schools follow the curriculum and promoting a culture of high expectations
Prioritize linking schools with parents and communities	Parents of SAR tend to be less involved in their children's schooling	Prioritize their links with parents and communities. Improve their communication strategies. Working with communities to enhance learning in schools

The results are from the study carried out over four years in a school in Israel on all the students enrolled in the IT course. They show increased interest of students at risk for learning by the increasing number of students that are using blogs for their education. As presented in the research, in 2013 there were 40 blogs uses, 44 in 2014, 58 in 2015 and 65 in 2016. The number of students at risk who dropped out studies is: 6 in 2013, 6 in 2014, 5 in 2015 and 2 in 2016. A blog uses per student statistic taking into account all the students participating in the study shows an increasing trend, as follows: 1.8 uses in 2013, 1.9 uses in 2014, 3.4 uses in 2015 and 5 uses in 2016. Considering that the number of drop-out students is decreasing and their blog uses is very low, we can neglect these values and get the following statistics of uses per student: 2.5 uses in 2013, 2.6 uses in 2014, 4.8 uses in 2015 and 5.9 uses in 2016.

In order to have better and more rigorous metrics for the degree of learning efficiency and the degree of learner engagement and satisfaction we are implementing this year several Google Analytics tools and gamification technics. Also, the research will be extended to other similar schools.

References

1. Mikkelsen, J., Holm, H.A.: Contextual learning to improve health care and patient safety. Educ. Health **20**, 3 (2007)
2. Argyris, C., Schon, D.: Organizational Learning: A Theory of Action Perspective. Addison-Wesley Publishing Co., Reading (1978)
3. Palmer, D.: What is the best way to motivate students in science? Teach. Sci. J. Aust. Sci. Teach. Assoc. **53**(1), 38–42 (2007)

4. Madrazo, D.: The effect of technology infusion on at-risk high school students motivation to learn. A Ph.D. Dissertation in Education. Reich College of Education (2011)

5. Barley, Z., Lauer, P.A., Arens, S.A., Apthorp, H.S., Englert, K.S., Snow, D., Akiba, M.: Helping at-risk students meet standards: a synthesis of evidence-based classroom practices. Mid-continent Research for Education and Learning, Aurora, CO (2002)

6. Darling-Hammond, L., Zielezinski, M., Goldman, S.: Using technology to support at-risk students' learning. Stanford Cent. Oppor. policy Educ. (2014)

7. McGuinnes, M.: Technology-based literacy instruction for at-risk students. J. Cross-Discip. Perspect. Educ. **8**(1), 12–20 (2015)

8. Jimoyiannis, A., Angelaina, S.: Towards an analysis framework for investigating students' engagement and learning in educational blogs. J. Comput. Assist. Learn. **28**(3), 222–234 (2012)

9. Sawmiller, A.: Classroom blogging: what is the role in science learning? Clear. House J. Educ. Strat. Issues Ideas **83**(2), 44–48 (2010)

10. Bandura, A.: Self-efficacy: toward a unifying theory of behavioral changes. Psychol. Rev. **1977**(84), 191–215 (1977)

11. Ashton, P.: Teacher efficacy: a motivational paradigm for effective teacher education. J. Teach. Educ. **35**(5), 28–32 (1984)

12. Santrock, J.: Educational Psychology: A Tool for Effective Teaching. McGraw-Hill, New York (2009)

13. Butler, Y.: Parental factors in children's motivation for learning english: a case in China. Res. Papers Educ. **30**(2), 164–191 (2015)

14. Chen, C., Chou, M.: Enhancing middle school students' scientific learning and motivation through agent-based learning. J. Comput. Assist. Learn. **31**(5), 481–492 (2015)

15. Larson, L., Kuhn, C., Collins, R., Balthazor, G., Ribble, M., Miller, T.: Technology instruction: fixing the disconnect. Princ. Leadersh. **10**(4), 54–58 (2010)

16. Eyvind, E., Knut-Andreas, C.: Perceptions of digital competency among student teachers: contributing to the development of student teachers' instructional self-efficacy in technology-rich classrooms. Educ. Sci. **7**, 27 (2017)

17. Lawlor, J., Marshall, K., Tangney, B.: BRIDGE21–exploring the potential to foster intrinsic student motivation through a team-based, technology-mediated learning model. Technol. Pedagogy Educ. **25**(2), 187–206 (2016)

18. Gan, S.: Motivating at-risk students through computer-based cooperative learning activities. Educ. Horiz. **77**(3), 151–156 (1999)

19. Lin, H., Fan, W., Wallace, L.: An empirical study of web-based knowledge community success. In: Proceedings of the 40th Hawaii International Conference on System Sciences, pp. 1530–1605 (2007)

20. Tekinarslan, E.: Reflections on effects of blogging on students' achievement and knowledge acquisition in issues of instructional technology. Int. J. Instruct. Technol. Dist. Learn. (2010)

21. Karademir, T., Oztürk, T.H., Yilmaz, G.K., Yilmaz, R.: Contribution of using e-portfolio system with peer and individual enhancing computer skills of students. In: Chamblee, G., Langub, L. (eds.) Proceedings of Society for Information Technology and Teacher Education International Conference 2016, pp. 936–941 (2016)

22. McBride, J., Henley, J., Grymes, J., Williams, D.: Effectively organizing and managing an electronic portfolio. In: Proceedings of Society for Information Technology Teacher Education International Conference 2015, pp. 975–980 (2015)

23. Theodosiadou, D., Konstantinidis, A.: Introducing e-portfolio use to primary school pupils: response, benefits and challenges. J. Inf. Technol. Educ. Innov. Pract. **14**(1), 17–38 (2015)

Environmental Monitoring Systems in Schools' Proximity Areas

Simona Riurean[(⊠)], Sebastian Rosca, Cosmin Rus, Monica Leba, and Andreea Ionica

Department of Computer and Electrical Engineering, University of Petrosani, Universitatii Str. 20, 332006 Petrosani, Romania
sriurean@yahoo.com

Abstract. This paper presents two research directions, both of them centered on the new available communication technologies that support Smart Cities concept applied into the crowded urban areas that aim to solve the issues of human health, especially for children. The automated monitoring of some of the main environmental parameters as well as data acquiring and analysis are planned to be done for a certain period of time in the schools' proximity in Petrosani, part of the Jiu Valley area, one of the Romanian most crowded coal mining basin not long time ago. Both the monitoring system, data acquisition and the related hardware, software and communication support as well as the noise control mapping near some major high schools in our community, are presented in the paper.

Keywords: Smart Cities · Noise map · Environmental health parameters

1 Introduction

The main challenge of a smart city is to be able to develop and effectively incorporate the new "digital space of the city" with solutions and tools so as to achieve the municipal authorization, which is the real objective of the smart city. It is a reality made not only of roads, buildings and physical infrastructure, but a city connected, networked, where the measure of "smartness" is given by a different insight of quality of life, to which population voluntarily contribute with their daily activities.

Nowadays, we are forced to face the multiple effects of the polluted environmental impact on human health and personal performance caused especially by the worldwide technological advance. In this circumstance, we consider important to use these advance technological state of the art means in order to monitor, collect and transmit data regarding important environmental parameters that have direct undesirable effects on population and especially on children. Based on these collected data improvements in daily children's life is possible to be done with municipal authorities and community' support.

In order to decrease or, in some cases, avoid negative effects caused by pollution of any kind (noise, unappropriated temperatures, humidity and so on) we are about to implement automated monitoring systems in some educational facilities in Petrosani

© Springer International Publishing AG, part of Springer Nature 2018
T. Antipova and Á. Rocha (Eds.): MosITS 2017, AISC 724, pp. 47–55, 2018.
https://doi.org/10.1007/978-3-319-74980-8_5

town, part of the Jiu Valley region. Additionally, is both useful for our children's health and also educative to draw up a noise map around some high schools in Petrosani.

1.1 Environmental Main Parameters and Optimum Level Regulation

International Organization for Standardization (ISO), European Committee for Standardization (CEN) and American Society of Heating, Refrigerating and Air Conditioning Engineers (ASHRAE) are writing standards relating to the indoor environment on a regular basis.

The environmental parameters that constitute the thermal environment are: temperature (air, radiant, surface), humidity, air velocity as well as personal parameters depending on clothing and specific activities. Criteria for an acceptable thermal climate are specified as requirements for general thermal comfort related to the air temperature and mean radiant temperature, air velocity, humidity and ambient light [1].

A continuous monitoring of all these parameters in schools is welcomed taking into consideration our permanent concerning for our children health and proper growth.

The recommended criteria for thermal comfort and ventilation rates according to CR (1752) (1998) and ISO/DIS 7730 (2003), for classrooms are as follows [1]:

- temperature level 24.5 °C during summer and 22.0 °C during winter;
- maximum mean air velocity 0.18 m/s during summer and 0.15 m/s during winter;
- basic ventilation 6 l/s m^2.
- humidity between 45% and 55%.

Knowing the mean average of the optimal environmental parameters for children in class, as well as in the school yards during brake, we will define the areas with problems that can affect children health and their everyday comfort. As a result of the research the proper measure has to be taken in order to maintain the environmental parameters in school at their optimal levels.

1.2 Noise Level Regulation and Effects on Humans' Health

Although not consciously, one of the main sources of irritation for children is environmental noise, caused by external factors and conducting to health problems and also to natural ecosystem alteration. This type of pollution occurs in urban areas, often leading to physiological and psychological injuries that are often detected when the damage has already been done. Environmental noise is the one produced by various human activities, such as transport, industry, to which the population is exposed, according to Environmental Noise Directive 2002/49/EC. This EU Directive rules in Europe on the assessment and management of environmental noise. This is an important legislative instrument for protecting citizens from excessive noise pollution [2].

It has been proven due to many scientific measurements and result analyses that prolonged exposure to high levels of noise pollution can have serious health effects. The endocrine disorder, cardiovascular disease, sleep disorders and discomfort are some of the direct effects on children's health as a consequence of long exposing on high level of noise. In Europe, according to the World Health Organization (WHO), after air pollution, the noise generates a high level of morbidity [3]. In the European

Union, approximately 40% of the population is exposed to road traffic noise with an equivalent noise level above 55 dBA in the daytime period and 20% of the population is exposed to levels above 65 dBA [4].

2   The Hardware Monitoring Support

The system used here is based on the low-power, high performance simple link wireless microcontroller that has 75% lower power consumption than Bluetooth smartphones in previous years, being powered from a long-life coin-operated battery. The wireless MCU (microcontroller unit) targets Bluetooth Smart, ZigBee and 6LoWPAN remote control applications. The RF device is cost-effective, ultra-low power on 2.4 GHz. Due to the fact that the integrated radio frequency transmitter and MCU needs very low currents and low-power consumption, an excellent battery life is provided. This ability also lets the device operate on small coin cell batteries and in energy-harvesting applications.

Thanks to the smart Bluetooth module in the Simple Link Sensor Tag kit, data from the sensors can be accessed online in less than 3 min via a cloud connection. It can be accessed from both iOS operating system smartphone and Android operating system by installing the dedicated application *BLE SensorTag*. It can also be accessed from a PC, through dedicated Sensor Tag Reader.

The sensor tag Bluetooth module uses the iBeacon protocol. This protocol allows the smartphone to launch applications and personalize content based on Sensor Tag data and physical location data. It is also compatible with ZigBee protocols (the mash wireless protocol with a low data date, low power consumption, and lower cost of automation and remote control) and 6LoWPAN.

The main features of this system are:

- offers low power;
- cloud connectivity, meaning that the sensors can be accessed and control from anywhere allowing to explore a seamless integration with mobile applications and web pages through JavaScript and MQTT. MQTT is a machine-to-machine (M2M) Internet of Things (IoT) connectivity protocol being ideal for mobile applications because of its small size, low power usage, minimized data packets, and efficient distribution of information to one or many receivers;
- supports multi-standard wireless MCU;
- it supports 10 low-power sensors (ambient light sensor, infrared temperature sensor, ambient temperature, accelerometer, gyroscope, magnetometer, pressure, humidity, microphone, magnetic sensor) [5].

Infrared thermopile temperature sensor measures the temperature of an object without direct contact with it. The integrated thermopile absorbs the infrared energy from the object in the field of view of the sensor. The device digitizes the thermopile voltage and then provides it and the die temperature as inputs to the integrated math engine. The math engine then computes the temperature of the corresponding object.

The humidity sensor with integrated temperature sensor is a factory-calibrated digital sensor with an integrated temperature sensor that provides accurate measurements at very low power. It measures humidity based on a novel capacitive sensor and functions within the temperature range of –40 °C to 125 °C. The innovative package device protects against dirt, dust, and other contaminants.

The ambient light sensor measures the intensity of visible light. The spectral response of the sensor closely matches the photonic response of the human eye and includes infrared rejection.

The digital humidity sensor provides excellent measurement accuracy at very low power. The device measures humidity based on a novel capacitive being factory calibrated. The sensing element of it is placed on the bottom part of the device, which makes it more robust against dirt, dust, and other environmental contaminants. It is functional within the full −40 °C to +125 °C temperature range (Fig. 1).

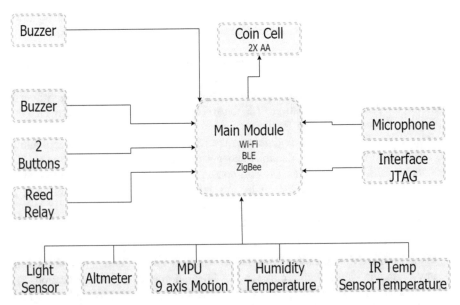

Fig. 1. The block diagram of the system shows the 10 low power sensors: ambient light sensor, infrared temperature sensor, ambient temperature, accelerometer, gyroscope, magnetometer, pressure, humidity, microphone, magnetic sensor [6].

The sensor controller is an autonomous processor that controls the peripherals independently of the main CPU that stays in the standby mode or runs a numerical ADC analog conversion probe or interrogate a digital sensor through the SPI interface. The primary CPU saves both the energy and the time needed for activation that would otherwise be missed.

The sensor controller studio software allows configuration of the sensor controller and the peripherals aimed to control as well as the conditions in which the main CPU is enabled.

Both BLE transmitter and receiver are listening on the same channel broadcasting one hundred packets/second listening on the channel that is twenty MHz away. The average rate is 45.8 bps being able to operate up to 4.4 m [7].

3   The Software Monitoring Support

3.1   IAR Embedded Workbench

IAR Embedded Workbench IDE Development Environment used here provides a very powerful collection of toolkits that offer extensive debugging capabilities, complex code execution, data breakpoints, stack execution analysis, stack view, and monitoring energy consumption as well as a range of integrated accessory tools for static analysis and code execution analysis.

3.2   Smart RFTM Flash Programmer 2

This software is used to program flash memory, update the software and the bootloader through the CC-Dev Pack-Debug troubleshooting module of wireless microcontrollers with radio transmitting receivers, based on ARM (Advanced RISC Machine; RISC - reduced instruction set computing) hardware architecture.

The Smart RF Flash Programmer 2 software includes a user interface and a command line interface.

The main features of the Smart RF Flash Programmer 2 software are:

- programming of flash images in wireless microcontrollers provided with radio transmitting receivers;
- programming/updating the firmware and bootloader of the assessment kits provided with the microcontroller via the port;
- add a software image to existing software on your device;
- reading a device software image in binary, hex or ELF (Executable and Linkable Format) files (ELF and bin format for hardware-based ARM devices);
- verify the existence of the software image within the file on the device;
- programming flash lock bits;
- read/write MAC addresses (IEEE standard EUI64/48).

3.3   The Dev Pack Watch Module

All the parameters described above are also displayed in real time on a high-resolution polymer networked liquid crystal display LCD with its own memory of 96 × 96 pixel active matrix that connects to the Sensor Tag kit via Dev Pack connector. The Sharp LCD delivers a fluent motion graphics display with a 50% reflection and a low power consumption of 12 μW (Figs. 2 and 3).

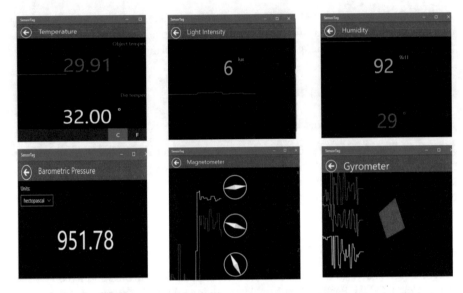

Fig. 2. The Sensor Tag PC interface for monitoring environmental parameters

Fig. 3. Parameters displayed on LCD watch Dev Pack

4 Noise Level Monitoring System

The Noise Level Monitoring System (NLMS) consists of a mobile application NoiseTube, a page on the social network Facebook as well as a Web application (Fig. 4). The mobile application integrates the noise levels measured by dedicate sensors; the social network helps to promote the issue around the theme and the Web application, based on Google Maps, allowing the noise levels map to be visualized.

Since noise pollution is a serious problem in many urban areas, the Noise Tube has been a research project started in 2008 at the Sony Computer Science Lab in Paris and maintained by the Software Languages Lab at the Vrije Universiteit Brussel [8]. The NoiseTube project proposed a participative approach for monitoring noise pollution by

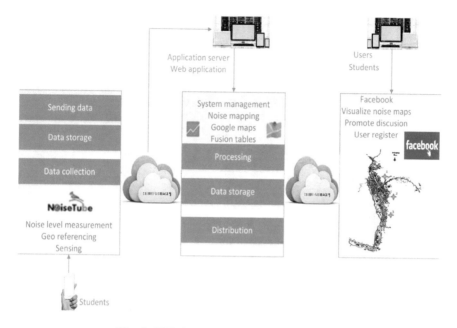

Fig. 4. NLMS: mobile application *Noise Tube*

involving the general public. The application aims a new usage of mobile phones by turning them into noise sensors enabling citizens to measure the sound exposure in their everyday environment. NoiseTube is an application that transforms the smartphone into an acoustic detector to measure noise pollution in a precise place and at a right time. Furthermore, each contributor in the project can participate in creating a collective map of noise pollution by sharing geo-localized measurement data with the Noise Tube community.

Noise Tube is built as an open source license, thus the source code has been published on Google Code [9].

The free application installed on the smartphone permits to measure the level of noise in dB(A) and tag the measurements obtained (e.g. subjective level of annoyance, source of sound, and so on).

When uploaded to the website (3G, Wi-Fi or manually) can be visualized the "sound trajectory" of the recorded measurements on Google Maps. The mediated map created based on the data collected from participants, draw the noise in different colors based on the decibel detected. Moreover, if the acquisition is carried out in a systematic manner, meaning that many children who repeat the same path several times (at least 50 takeovers), it will be possible to create maps with statistically reliable values [10].

To get accurate measurements, the software needs to be calibrated for each phone model (because of hard and software differences), mostly by downloading the calibration settings from the Noise Tube website.

The noise mapping process targets the real situation to be revealed and be available on the map to see daily exposure resulting from the presence of noise sources such as

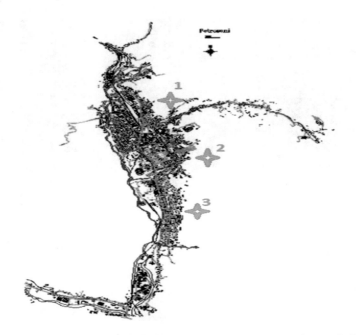

Fig. 5. Petrosani map with the 3 high schools monitoring project marked

commercial areas, railways, highways and industrial units in cities. This way, the map highlights the spatial variation of ambient noise levels (Fig. 5).

5 Conclusions

The system developed by us aims to perform a complex analysis of the environmental parameters in real time inside schools as well as at schools' proximity areas in order to improve both students' and teachers' comfort. In order to achieve what we proposed, we combined on the one hand the multi-sensor wearable devices technology based on the Internet of Things and smart Bluetooth with low power consumption for high-precision detection and display of environmental parameters on a display LCD attached to the SensorTag and a PC application on time slots, and on the other hand smartphone application technology for detecting background noise in the school's perimeter. Based on the data acquired, a daily and weekly offline analysis can be performed. In order to increase the accuracy of the data received, a significant number of students are involved in our research who installed their noise detection application on their own smartphones and who are active while they were going to school.

As a result of the data processing acquired through this widely implemented system, it is possible to optimize the cost of school warming due to high-precision measurements of environmental parameters that can be extended beyond the classroom hours.

Major technological advances in recent years also bring a major impact on the environment and thus on users. That is why we need to develop a balance between the benefits of technology and the health of users. This can only be done by conducting environmental impact studies of relevant technologies (automotive industry, IT technologies), and the most appropriate place for such experiments would be the proximity to educational units, as here both the technologies of a small smart city and a relatively large mass of users coexist.

Next generations must benefit both from new technological achievements and the best of live in an optimal environment to carry on the technological advance. Smart City is not about technology, smart city is about people and the improved interconnections between them for a better life.

References

1. Olesen, B.W.: International standards for the indoor environment. Int. J. Indoor Environ. Health **14**(s7), 18–26 (2004)
2. Heutschi, K.: Lecture Notes on Acoustics I. Swiss Federal Institute of Technology, ETH, Institute for Signal - and Information processing, ISI, Zurich (2013)
3. Goelzer, B., Hansen, C., Sehrndt, G.: Occupational exposure to noise: evaluation, prevention and control. World Health Organisation, Geneva (2001)
4. EPA: Guidance Note for Strategic Noise Mapping. Environmental Protection Agency, Ireland (2011)
5. Shibata, T., et al.: Energy consumption and execution time characterization for the SensorTag IoT platform. In: XXXIV Simpsio Brasileiro de Telecomunicaes e Processamento de Sinais (SBrT 2016), pp. 55–59. Santarém, Brazil (2016)
6. Texas Instruments Multi-Standard Sensor Tag Design Guide. http://www.ti.com/tool/SENSORTAG-SW#technicaldocuments
7. Zhang, P., et al.: Enabling practical backscatter communication for on-body sensors. In: Proceedings of the 2016 Conference on ACM SIGCOMM 2016 Conference, pp. 370–383. ACM (2016)
8. Thenkabail, P.S.: (ed.). Remotely Sensed Data Characterization, Classification, and Accuracies, pp. 435–436. CRC Press, Boca Raton (2015)
9. Noise Tube Application. http://www.noisetube.net
10. Rey Gozalo, G., Barrigón Morillas, J.: Analysis of sampling methodologies for noise pollution assessment and the impact on the population. Int. J. Environ. Res. Public Health **13**(5), 490 (2016)

Designing a Public Sector Accounting Exercise Sheet by a Cultural Approach

Indra Bastian[(✉)]

Accounting Department, Gadjah Mada University, Yogyakarta, Indonesia
indra_bastian@yahoo.com

Abstract. Teaching a public sector in an undergraduate class was challenged. Many traps on Many materials have been investigated as a number feature without interpretation. In the end, the number became a fixed and limited form. In this case, the number should be reflected as a symbol and a condition of behind and after an event.

To understand what the event is, we need to observe and explore a reality around a public sector office. In fact, a connectivity among public offices under one area is a reality attached in how activity can be traced. Besides that, many procedures from above governments are comply as main executing procedures.

Both main conditions in a public office can be interpreted as a value and tradition inside of a public office. So, a cultural approach can be utilized in a design of a technical sheet.

1 Teaching by a Simulation

One of the most important steps in developing curriculum is the introduction of learning based on simulation. Simulation is a generic term referred on artificial representation of real world process to achieve educational goals through experience learning/experiential (Al-Elq 2010).

According to Language Center of Ministry of Education (in Betty Sufiah et al. 2014) simulation is a training method that demonstrate something in the imitation form that similar with the real situation; simulation is depiction of a system or process of demonstration by using statistic model or cast. Simulation also can be defined as a replica or visualization of the behavior of a system that is running in a certain period of time (Sa'ud in Betty Sufiah et al. 2014). On other words, simulation is a model that consists of a set of variable that present a main features from the real life system. Simulation enables to help party involved to modify the decisions that determining how the main features in real.

When the students use behavior model to get a better understanding on the behavior, they are doing a simulation. For example:

- When the students are assigned roles as buyers and sellers of something and asked to deals to exchange the good, they are learning about market behavior by simulating a market.

T. Antipova and Á. Rocha (Eds.): MosITS 2017, AISC 724, pp. 56–67, 2018.
https://doi.org/10.1007/978-3-319-74980-8_6

- When the students take on the roles of party delegates to a political convention and run the model convention, they are learning about the election process by simulating a political convention.
- When the students create an electric circuit with an online program, they are learning about physics theory by simulating physical set-up in an actual.

Teaching by a simulation or instructional simulation have the potential to engage students in "deep learning" that empowers understanding as opposed to "surface learning" that requires only memorization. Deep learning means that students (Blecha and Maier 2017):

1. Learn scientific method includes:

   - the importance of model building.
   - the relationships among variables in a model or models.
   - issues of the data, probability and sampling theory.
   - how to use a model to predict outcomes.

2. Learn to reflect and extend the knowledge by:

   - actively engaging in among students or instructor-student conversations needed to conduct a simulation.
   - transforming knowledge to new problems and situations.
   - understanding and completing through their own processes.
   - seeing social processes and social interactions in action.

Furthermore, simulation method also aims to (Betty Sufiah et al. 2014):

1. train certain skills in professional or for daily life;
2. obtain understanding on a concept or principle;
3. train to solve problems;
4. improve the learning liveliness;
5. give learning motivation to the students;
6. train the students to cooperative in a group situation;
7. foster the creativity power of the students, and;
8. train the students to develop tolerance attitude.

Teaching by a simulation is very helpful to create a prediction on conditions of social, economics, or the world naturally. Therefore, teaching by a simulation also suitable to learn on things related to public sector accounting. There are several types of simulation model that can be used to help students to be more understanding public sector. Types of simulation method, among others (Betty Sufiah et al. 2014):

1. Role playing

   In the learning process, this method gives priority in the form of dramatization. Dramatization is done by a group of students with implementation mechanism directed by lecturer to implement the activities that have been determined or planned previously. This simulation is more emphasis on the goals to recall or re-create a picture of the past which is possible to happen in the future or an actual event and meaningful for the current life.

2. Sociodrama
   Sociodrama is a learning method of role-playing to solve problems associated with social phenomenon, problems that associated with the relationship among human beings.
3. Simulation Games
   In its learning, students are role-playing according to their assigned roles as learning to make a decision.
4. Peer Teaching
   Peer Teaching is a teaching practice done by students to the prospective teachers.

Any type of simulation method that is used to teaching, need to be noted that to achieve an effective teaching with simulation method requires (Blecha 2017):

- Preparation of instructor or lecturer or teacher. Instructional simulations or teaching by a simulation can be very effective in stimulating student understanding, however, it requires intensive lesson preparation.
- Active student participation. The learning effectiveness of instructional simulation rests on actively engaging students in problem solving.
- Post-simulation discussion. Students need sufficient time to reflect on the results of simulation.

2   The Using of Simulation for Public Sector Accounting

Simulation method is very suitable to be used in learning based on contextual, the learning material can be based on social life, social values, or social problems.

Since the early 1960s a number of business simulations have been developed in financial management, corporate governance, marketing and other fields. This effort is mainly done to improve the quality of the learning process and to improve the application of the skills and techniques gained to actual managerial practices.

Nevertheless, the experience of business simulations has not been widely transferred to teaching for public administration and similar programs to train managers in the Public Sector. Issues of improvement in action training and future representation of public sector governance have not been widely discussed.

Volkov et al. (2004) conduct a special research aimed at detecting the presence and use of simulations in public administration programs (which in this case relate to public financial management and public sector accounting). The study includes top universities in Russia as well as a number of Western institutions that offer a variety of simulation products. There are no simulated products in public finance found by this research. Education in public administration and equivalent programs along with the practice itself is far more conservative than education and practice in business administration. In the business sector, driven by competition, change occurs quickly compared to the public sector where changes are very slow and rare. Even in transition countries, such as Russia, the pace of change of "state machines" is significantly slower than in the business sector. The same rule is concerned in business education compared to civil service education (managers in public administration). The main disadvantages of existing education in public financial management in particular and public administration in general, at least in Russia, are:

- Long-term teaching (traditional learning form such as lecturing is very time-consuming);
- Knowledge-based orientation, not skill-based orientation;
- Scope of practical management decisions being divided into pieces and separated among a large number of different subjects/courses;
- Dominance of "listening", not "practicing";
- Rare in "update" of educational programs "content";
- Lack of available experts and qualified specialist in this area.

In teaching a simulation public sector accounting will help deeper understanding, and eventually students are expected to "gain experience" and later they can implement accounting practices in the public sector better. Students are expected to be trained to see the relationships among variables in the model and to transform knowledge to new problems and situations until finally getting used to making the right decisions about the problem.

Teaching by a simulation also allows exploring the reality that inherent of any public finance governance activity. One of them is culture, where the accounting system is not a subject that is only influenced by culture but by all elements of the economic, political and legal background in which it is applied. The accounting system is simultaneously influenced by the background in which the entity operates and the organizational culture of each unit.

Culture has an impact on decision-making. A very interesting context to learn is decision making through cross-cultural teams, where decision-makers have different cultural backgrounds. One of the important aspects of cross-cultural analysis is the division between western and non-western cultures. It has been shown that representatives of the western world have a more individualist that is more individual perception of self; On the other hand, those from non-western worlds often choose collectivist consciousness. The difference of this type has the potential to produce different types of decision-making strategies, which may prefer individualist or collective orientation. In these circumstances, one important question is whether these different decision-making strategies are reflected in decision-making behavior, and, if so, how they can be learned?

With simulation-based teaching, exploration of different decision-making strategies can be made in the decision-making environment. Basically, the data in the context of public sector accounting is generated from the decisions of the simulation game participants, which are recorded into the database during the game simulation sessions. This database has several attributes such as time stamps, which indicate certain decisions are made in simultaneously with the type of decision. This information is then possible to be associated with the demographic and socioeconomic factors of the decision-making group and also the different performance indicators of the decision group's simulations. Such data allow for a more realistic decision-making analysis and consider factors such as national culture. As a simulated reality, rich game data can open an entirely new and unpatented environment for researching decision-making; It also has the potential to provide insight, which can greatly improve our understanding of decision making in the field of public sector accounting (Kallio 2015).

3 Making a Working Sheets Design for Simulation Working Paper

To explain the design of simulation working sheets, the writer will take an example of public finance simulation design in Russia.

Basis for simulation development is the principle of budgeting system. The Russian budgetary system is very young and changing dynamically. Since the collapse of centralized economy and unified budgetary system in 1991, there have been several stages of budgetary reforms. At this moment, the Russian budgetary system almost completely reflects the way the national economy is organized, yet it is still being modernized and experienced many changes. Due to its history, there is a significant difference between the budgetary system of Russia and other federal countries such as USA, Germany or Brazil, in terms of powers and responsibilities between budgetary system levels, structure of taxation system and public finance management instruments used. However, as in every country built on federal principles, there are three levels in the budgetary system of Russia, they are (Volkov et al. 2004):

- One Federal budget (central government);
- 89 Regional budgets (similar to State budgets in USA);
- More than 29000 local (or municipal) budgets.

As a result, different authorities are responsible for the federal, regional, and municipal budget processes (Fig. 1). In addition, the rest of the Soviet Union are the four major public infrastructures that are mostly in the area of public financial responsibility, and therefore largely are financed by the budget system, and remain part of the state's wealth. These areas are education, public health services, social security, and housing and communal services. Currently, even under the new education system, medical services and social security are mostly provided free of charge by state organizations (budget) and remain the responsibility of the state. Basically, the system operates as follows: businesses pay different taxes and fees for different budgets and some non-budgetary funds and these resources are used to finance public infrastructure. The fact that some of the most important functions of this system included in regional budget are another reason for developing effective training technology for the regional budget system.

Simulation of regional public finance management (Fig. 2) is based on the general principle of public finance management at the regional level relating to federal states (countries). Moreover, the model strongly reflects the structure of the budget system in the Russian Federation. There are instruments in the Russian budget sector that are different from those used in the United States and other countries, therefore, it is sometimes difficult to find the right words to discuss some simulation problems and the Russian budgetary system.

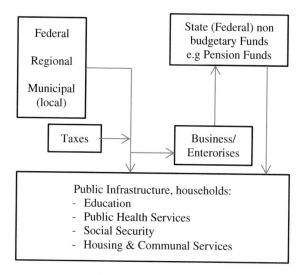

Fig. 1. Basic structure of Russian budgetary system

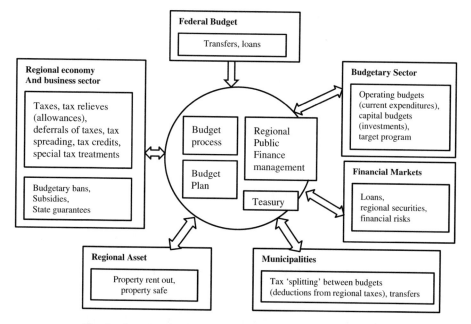

Fig. 2. Aspects of public finance management on regional level

The simulation is aimed to develop the skills and techniques that relate to:

- Managing budgetary revenues, tax administration, regional economy regulation and stimulation;

- Managing current budgetary expenditures (operational budgets) and capital investments (capital budgets);
- Determining sources of budget deficit coverage and managing regional debt;
- Managing regional assets (specifically, state property) in collection of non-tax revenues;
- Developing the relationship between the Laws and regulations on the regional level.

The following particular cases and situations of public finance management are simulated in the game.

Budgetary revenue, tax administration, the state support on regional economy:

- estimating tax revenues for next financial year (fiscal);
- analyzing regional budget arrears;
- deciding regional tax rates;
- deciding relieves to different industries;
- deciding deferrals of taxes and tax credits;
- deciding budgetary loans and guarantees;
- deciding special tax treatment for small businesses.

Administration and control on the budgetary expenditures:

- planning current expenditures (operational budgeting);
- planning capital investments (capital budgeting);
- determining principles and "packages" plan of target programs financing (priority financing of specific industries);
- determining principles and planned amounts for specific capital projects;
- analyzing possible deviations from planned balanced budget;
- deciding "protected items" (financed in full state) of budget;
- deciding and revising amounts of financing on budget execution stage.

Management of regional debt:

- formulating and servicing direct debt portfolio;
- accounting and servicing contingent (indirect) debt portfolio;
- considering seasonal factors in revenues collection and covering temporary cash deficits (gaps);
- estimating influence of financial markets condition on possibilities and terms of loan funds;
- managing accounts payable (for purchased goods and services).

Management of regional property:

- setting standards of rental payments and amounts of regional state property sales;
- sale of regional property (assets). Inter-budgetary relations:
- determining and setting standard values of deductions from regional taxes to municipal/regency budgets;
- determining the size (amount) of the fund of financial support to municipalities/regencies (FFSM);
- determining the proportional shares of FFSM distribution (aligning provisions and incentive provisions).

The participating teams are responsible for managing public finances of virtual regions and acting as leaders of the regional government finance department (Fig. 3). At the beginning of the game, all regions have the same financial and economic conditions. As a result of applying team strategy in the game process, the financial and economic conditions of the region can greatly vary. Along with short-term operational decisions, teams should carefully consider long-term capital investment strategies into the development of public infrastructure. Such a policy can help achieve the best value of the target indicator. The team acts to consider changing economic situations and federal norms and standards set by game providers through scenarios and regulations. Teams compete with each other for:

- attraction of external investments into regional economy;
- attraction of labor;
- regional markets of goods and services for businesses operating in regions.

Team performance measures and the ranking system are based on current and estimated values of "target indicators".

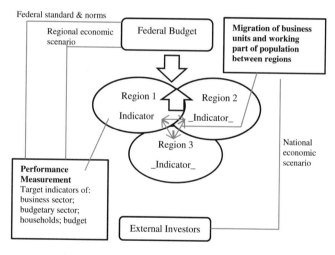

Fig. 3. Structure of the simulation

4   Filling the Working Sheets with Real Data: Interpretation Challenges

Simulations are based on complex computer models that process team decisions, calculate results, and make evaluations using a set of specific parameters (Fig. 4). A large number of parameter changes are incorporated into game model scenarios. The game scenario combines different situations and cases similar to the public finance managers face in their actual practice. They are prepared and put into computer models by game developers. To improve the diversity of game situations and cases, random factors were introduced on computer models.

The team made their decisions based on about 40 reports reflecting the conditions of economic and social sectors, regional budgets and debt, regional state property, municipal finances and so on. Team decisions are incorporated into a specific decision-making forms. According to the game schedule, this form is forwarded to the instructor and entered into the computer program. From special instructor reports, game organizers can monitor the team decisions and change the limits for team decision scores, and also, change individual game rules. This activity helps to keep track of what is going on in the game and control the game process.

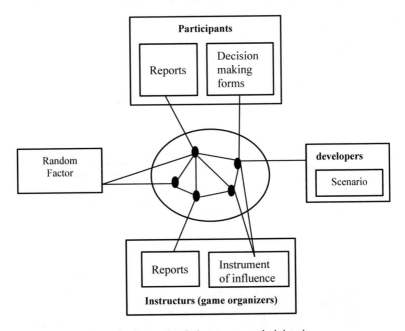

Fig. 4. Game simulation process administering

The simulated region is represented by:

- economy diversified by industries and territories which also includes a number of largest tax payers;
- budgetary sector (5 budget industries) and special State Regional Road Fund;
- complex of regional property;
- 10 municipalities - 4 urban municipalities and 6 rural municipalities mainly specialized on agriculture;
- all main groups of population.

The basic game scenario has a very powerful impact on the simulation process. Game scenarios are represented in the simulation as the changing dynamics (time set) of different regional economic, financial and budget indices. Estimating the possibility of scenario indices changes is one of the issues that the team must resolve. Medium-term social and economic development estimates distributed to the team each period before the budget preparation (planning) for the next financial year is a necessary source of information. Random factors increase the complexity of the simulation due to the formation of series of cases and events for each fiscal year (e.g. list of capital investment projects or regional target programs). Random events help make the simulation more realistic. Transfers from central government budgets in simulations serve to play the instrumental role of stabilizing the budgetary system and leveling differences in budgetary security between regions (Fig. 5).

One game period equals a quarter of a year. One game cycle equals four game periods or one fiscal (financial) year. During one fiscal year, the team conducts budget execution for the current fiscal year and budget preparation for the next fiscal year. The budget preparation for the next fiscal year is done in the last two quarters (two periods) of the previous year (currently) (Fig. 6). The team's decisions become the regional budget rules and are represented by the amount budgeted with the budget of the budget recipients shown.

The training session structure for the simulation is very flexible. Usually a short and intensive training session lasts five days. Usually, the first part of the first day team gets acquainted with the simulation instruments and the game process and asks questions about the model and various game procedures. During the general tutorials that follow the game's experimental period, the team analyzes how computer models respond to their decisions; then the actual period of the game begins. In the next three days the team was involved in intensive decision-making and discussions on various public financial

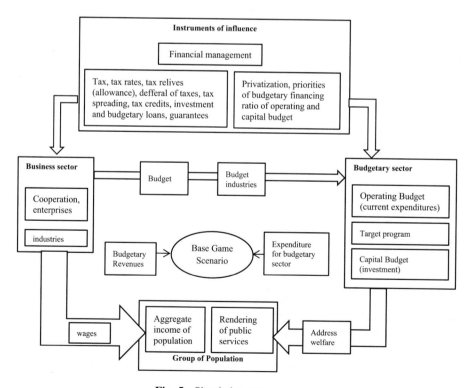

Fig. 5. Simulation game scenario

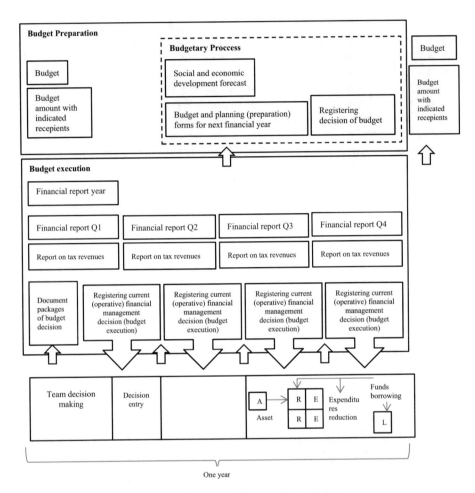

Fig. 6. Structure of Typical Game Period

management issues. During one day, the team made decisions for two fiscal years. The duration of the training session usually allows the team to prepare and implement 6 to 8 regional budgets (8 years equals two Governor elections in Russia). The last day is dedicated to the analysis of team activities carried out in the course of the game. The team analyzed the experiences they had gained in the last four days and made presentations about their strategies and activities to the panel of judges. The criteria for determining the winner is a set of special indicators that reflect regional economic conditions, budgetary sectors, budgets and debt, the amount of public services provided (services provided at budget costs). There are nearly 150 indicators calculated by computer models that reflect the major aspects of managing the regional tax base and public finance. The weighting factor used to determine the value of the final complex performance index (FCPI) is under the control of the game organizer and depends on the purpose of the training session. Winners are determined based on FCPI evaluating the quality of public finance management. In determining which winning instructors can also consider how teams carefully follow their strategies, how they organize their teamwork and how closely they analyze situations in the simulated regional budget system.

5   Conclusion: The Using of Culture Approach Framework in Making the Working Sheets of Public Sector Accounting

The use of culture as a metaphorical root arose because of the phenomenon of Japanese enterprise success in the 70 s and 80 s. Business leaders in America, while studying the methods of production in Japan, found that the quantity and quality of output of Japanese industrial enterprises was not caused by technology but because of the shared cultural values among their workers about loyalty to colleagues and their companies. Every organization is different. It depends on how people in group culture make meaning. Pacanowsky sees culture as more than a single variable in organizational research. According to him, organizational culture is not just a piece of the puzzle but the puzzle itself. In our view, culture is not something the organization has; culture is that organization.

In this case, a cultural approach is used to map the possible socioeconomic conditions and decision-making done by the team in the simulation game. Simulation-based teaching can reveal people who they are (students act as who), they think what they do, and what they think about their goals. There are no shortcuts to collect in-depth investigation reports about interactions. Without that clear understanding, nothing will be interpreted.

The simulation method can help to observe by becoming a member of a society. Nevertheless, he must be aware of his position as a researcher so that he is able to put the culture of the community he is examining as a source of knowledge held by the community. In addition, on why the community embraces that culture. Thick description requires interpretation and observation. Thick description traces the cultural network and reveals the meaning.

References

Al-Elq, A.H.: Simulation-based medical teaching and learning. J. Family Community Med. **17** (1), 35–40 (2010), https://www.ncbi.nlm.nih.gov/pmc/articles/PMC3195067/. Accessed 25 August 2017

Betty Sufiah, W., et al.: TEORI PEMBELAJARAN "Metode Simulasi" (2014), http://imadiklus. com/teori-pembelajaran-metode-simulasi/. Accessed 25 Aug 2017

Blecha, B., dan Maier, M.: Teaching with Simulations, 1 Agustus (2017), http://serc.carleton. edu/sp/library/simulations/index.html. Accessed 25 Aug 2017

Volkov, A.A., et al.: Teaching public finance management through simulation. J. Dev. Bus. Simul. Experiential Learn. **31** (2004), https://journals.tdl.org/absel/index.php/absel/article/ viewFile/675/644. Accessed 25 Aug 2017

Kallio, H.: Decision-making strategies in business simulation environment: a cultural approach **4** (6) (2015), http://aisel.aisnet.org/iris2015/4. Accessed 25 Aug 2017

Information Technology in Art

Exploring Open Innovation Technologies in Creative Industries: Systematic Review and Future Research Agenda

Narmina Rahimli[1(✉)] and Eric W. K. See-to[2]

[1] Department of Industrial and Systems Engineering,
The Hong Kong Polytechnic University, Hung Hom, Kowloon, Hong Kong
narmina.rahimli@connect.polyu.hk, narmina05@gmail.com
[2] Department of Computing and Decision Sciences, Lingnan University,
Tuen Mun, New Territories, Hong Kong
ericseeto@ln.edu.hk, eric.seeto@gmail.com

Abstract. The main objective of this paper is to analyse application of open innovation technologies in different sectors of creative industries, e.g. architecture or performing arts. This study adapts a systematic literature review of 29 papers in order to present the state of art, provide research gaps and set up directions for future research in the area. The results obtained highlight a growing interest to the subject area and at the same time demonstrate lack of empirical studies in creative sector. Originality of this study lies in providing a new angle to existing research on open innovation. Theoretically this paper serves as a starting point for future research. Practically it can be used as a guidance for creative managers.

Keywords: Open innovation · Open innovation technologies
Creative industries · Arts and humanities

1 Introduction

Innovations are vital for organisational success. In the past companies relied on innovations produced within organisational boundaries, i.e. R&D. However along with rapid advancements in information technologies in order to sustain the pace of innovation companies had to open up their business models to improve innovation processes and become more competitive. Open innovation is one of those concepts that encourages companies to open up business models in order to look for ideas outside of organisational boundaries to improve internal innovation processes and apply internal ideas to external processes, i.e. exploration and exploitation [1]. The concept of open innovation gained a momentum in 2003 when it was mentioned by Henry Chesbrough and since then it became one of the hottest topics in the area. Despite its popularity and provided benefits sources suggest that the adoption rate of open innovation concept is quite low among organisations [2].

Creative industries from other side is another example of how information technologies created and facilitated the emergency of new economy and provided demand

© Springer International Publishing AG, part of Springer Nature 2018
T. Antipova and Á. Rocha (Eds.): MosITS 2017, AISC 724, pp. 71–85, 2018.
https://doi.org/10.1007/978-3-319-74980-8_7

for new skills. Creative industries rose to prominence in 90s firstly in Australia, and later they become an important agenda for economic development when the government of Tony Blair set up the Department of Culture, Media and Sport (DCMS) to promote creative industries. Nowadays these industries play an important role in economic development. For example, creative sectors worldwide involve 1% of the world's economically active population [3]. Furthermore, creative industries represent a broad cluster with industries ranging from technological to non-technological sectors where knowledge is applied as an output and intellectual property produced as an output [4, 5].

Information technology plays an important role both in open innovation as an enabler for exploration and exploitation processes, and in creative industries due to their technological nature. However current research shows lack of sufficient studies that explore a phenomena of open innovation and information technologies in the context of creative industries.

Thus the aim of this paper is to analyse to what extent creative industries apply open innovation technologies. This study is structured around the following research questions: *(1) How do creative industries adopt open innovation technologies? (2) What are factors that influence the adoption of open innovation technologies in creative industries? (3) How do adopted open innovation technologies impact creative industries?* In order to address these questions this study applies a systematic literature review for in depth understanding of the current research domain.

The intended contributions of this study are the following. Firstly, this study addresses the existing literature gap in the field of open innovation which will be discussed in the next section. Secondly, this paper places the concept of open innovation in the context of creative industries to understand to what extent creative industries use innovative technologies. We believe that the findings and questions raised by this paper will be of interest to both academia and practitioners.

This paper is structured in four sections. The Sect. 2 provides a discussion on open innovation, open innovation technologies and its importance in the context of creative industries. This is followed by a methodology section that discusses systematic literature review and why it was chosen for this study. The Sect. 4 presents the results obtained through systematic review of 29 studies (21 journal papers, 8 proceedings). The Sect. 5 sums up the study, discusses limitations and implications.

2 Open Innovation, Open Innovation Technologies and Creative Industries

Since 2003 when Open Innovation was coined by Henry Chesbrough in his book *Open Innovation: The new imperative for creating and profiting from technology* it has become one of the most popular and cited topics in the field of innovation management. For example, the 2006 version of the book has been cited more than 13,000 times (Google Scholar, September 2017). Along with the progress in the field of innovation management, information technologies and growing interest to open innovation its definition has also undergone changes from original version: *"open innovation is a paradigm that assumes that firms can and should use external ideas as well as internal ideas, and internal and external paths to market, as firms look to advance their technology."* [6, p. xxiv] to more comprehensive: *"a distributed innovation process based on purposively managed knowledge flows across organizational boundaries, using pecuniary and non-pecuniary mechanisms in line with the organization's business model"* [7, p. 17). As it can be seen from the definitions provided as a research construct open innovation is built upon managing and advancing external and internal knowledge processes which can be also referred to as *inbound, outbound and coupled* activities [8]. Inbound activities refer to application of external knowledge to internal processes whereas outbound are defined as usage of internal knowledge for external purposes [9]. While "knowledge" perspective holds quite strong positions among researchers others believe that role of information technologies in open innovation research is overlooked as technology facilitates these internal and external knowledge processes [10]. So, what role does information technology play in open innovation processes and how researchers should approach it?

This question can be answered in two ways. Firstly, direct impact of information technologies on innovation processes. For example, Gordon et al., [11] conducted a study that analysed 80 companies where users were involved in the front-end innovation processes through information technologies. The results of this study showed that information technology positively facilitates users' involvement on collaboration, data gathering and knowledge generation, i.e. front-end innovation processes. Secondly, information technology serves as a platform for open innovation, i.e. open innovation technologies. Among open innovation technologies researchers distinguish idea management system, problem solving system, marketplace system and innovation analysis system [10]. These systems are built to promote and facilitate collaborations between users and organisations in a form of problems' solving, idea generation, evaluation of products and others [10]. In addition further research suggests that open innovation technologies are versatile by nature and range from brainstorming platforms to prototyping sandboxes, however managing these technologies can be complicated as its application requires changes on all levels of innovation processes [12]. Recent reviews and discussions on open innovation call to expand current research domain with inclusion of such technologies as innovation platforms, social media and crowdfunding [13]. Summing up these discussions and in the context of this study it would be reasonable to question how can open innovation be analysed in creative industries and why we should consider it as an important research direction?

Firstly, as already mentioned in introduction and according to multiple reports creative industries significantly contribute to economic growth of many countries and the numbers are yet to grow. For example, as of 2013 they generated combined 3% of the world's GDP which is equal to US$2,250b in revenue [3]. This can be explained by the fact that innovation is an important element of creative industries. Many scholars link creativity to innovation, which itself generates economic value [14–16]. Secondly, creative industries are represented by broad range of sectors from technological, e.g. software to non-technological, e.g. performing arts. Information technology plays an important role for creative industries as it lowers barriers to markets and facilitates better engagement with customers, stakeholders etc. For example, emergency of fast fashion like Zara which made well-established brands to become more competitive and recognise power of innovative technologies [17].

Thus in the context of this study open innovation in creative industries can be analysed from two perspectives. Firstly, open innovation in form of online platforms, e.g. digital arts platforms where users collaborate towards final product [18]. Secondly, direct impact of open innovation technologies on innovation processes in creative industries. For example, newly created platform TheCurrent that calls fashion brands to cut down on R&D and transaction costs through application of open innovation technologies [19]. Thirdly, because creative industries apply knowledge as an input and generate intellectual property as an output they may serve as a provider of external knowledge for large corporations. Combining these examples it can be seen that exploring the concept of open innovation in the context of creative industries creates an interesting research domain. However despite the popularity and importance of both open innovation and creative industries the research in this area is quite scarce.

The current research gaps in the field of open innovation show lack of studies across industries [1, 20, 21]; correlation of technologies used for open innovation and context they are applied [10]. In the context of creative industries at the moment research is still very new and emerging. Current research demonstrates ongoing debates on controversies surrounding definitions that should be used for creative industries [22, 23] and different perspectives on conceptual approaches, e.g. industry vs social market [24].

Despite the lack of sufficient academic research on this topic the importance of exploring and understanding of how open innovation/open innovation technologies are applied in creative industries has been a central topic in special issue by AMA (Open Innovation in Creative Industries) and various governmental foundations like NESTA. Therefore we believe that creative industries introduce a new and interesting research area because these industries not only rely on innovations and information technologies, but also produce economic value in a form of intellectual property as already discussed above. Secondly, by conducting the review on the current research topic this study will expand the open innovation research area and address some of the gaps mentioned above.

The Sect. 3 of this paper will provide an overview of methodology used followed by results obtained.

3   Methodology

In order to analyse and meet the objectives of this paper we apply a systematic literature review that consists of the following steps [25]: to plan the review, to conduct the review, and to report the review results. Such approach provides a detailed and transparent analysis based on a theoretical combination of existing research.

(1) *Planning*

This step involves identifying research focus. In the context of this paper the research focus was set on studies that deal with open innovation, open innovation technologies and information technologies in creative industries' context.

(2) *Conducting*

In order to extract papers relevant to the research context of this study we had to identify the list of relevant keywords. This process was split in the following stages. Firstly, search query *"open innovation" AND "creative industr*" OR "cultural industr*"* was generated and applied for search in Web of Science by Thomson Reuters which was chosen as a primary database for this paper. This search has yielded very few results. Therefore, the search strategy was expanded to include more keywords such as *"networked innovation"*, *"distributed innovation"*, *"collaborative innovation"*, *"crowdsourcing"*, *"crowdfunding"* in combination with *"creative industr*"* OR *"cultural industr*"*. Thirdly, once we extracted relevant papers through the previous search application the search query was expanded into finding the industry specific papers, e.g. open innovation technologies in visual arts. Thus, on this stage search query contained such keywords as *"open innovation"* OR *"networked innovation"* OR *"distributed innovation"* OR *"collaborative innovation"* OR *"crowdsourcing"* OR *"crowdfunding" AND "TV"*, *"art*"*, *"newspaper*"*, *"advertising"*, *"architecture"* etc. For industries selection we applied DCMS list which is internationally recognised by governments and institutions [5]. Therefore, articles dealing with a rising phenomena of open innovation technologies in science or food industries were excluded despite the fact that some scholars see it as a part of creative industries [e.g. 26].

(3) *Reporting*

Following the completion of steps 1 and 2 two-stage report was generated. The report firstly provides a descriptive analysis of results obtained, e.g. papers by sources. Secondly, the report presents an overview of content findings in line with the research objectives based on three subtopics identified.

Figure 1 showcases the papers' selection process through the PRISMA flowchart. The initial sample of papers obtained through databases' search was 83. On this stage twenty three papers were excluded based on the following criteria: full text availability, language and duplicates. Following this exclusion procedure 60 papers were selected for the 2nd round. On this stage in order to ensure the relevance of papers authors carefully read abstract and full-text of each paper. This screening processes downsized the number of papers from 60 to 29.

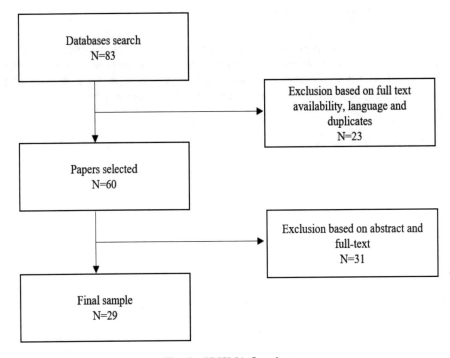

Fig. 1. PRISMA flowchart.

Upon the agreement on the final sample of papers between authors the two-stage report has been produced. The results obtained will be presented in the Sect. 4.

4   Findings and Discussion

4.1   Descriptive Statistics

This section presents descriptive analysis of 29 papers based on papers published over time, by journals, by methodology applied and by industry type.

4.1.1   Papers Over Time

Figure 2 shows distribution of selected papers over time. As it can be seen from the chart the highest number of papers (7) was published in 2017 which shows growing interest to the subject area. In fact the selected sample demonstrates the constant trend since 2015. In 2009 UNCTAD addressed Creative Economy as an agenda for economic development which also could explain the interest to the subject area starting from 2010 in the selected sample.

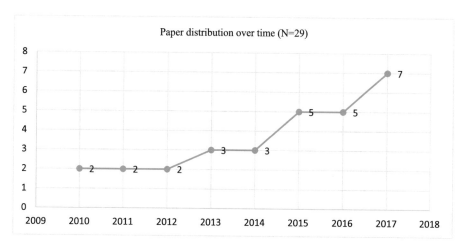

Fig. 2. Paper distribution over time.

4.1.2 Papers by Journals

Table 1 demonstrates the allocation of papers by journals with more than two publications. The average impact factor for journals showed in Table 1 is 1.51 with *Technological Forecasting and Social Change* having both the highest impact factor *(2.678)* and number of papers *(3)* published followed by two papers published in *International Journal of Technology Management* and two papers in *International Journal of Arts Management.*

As for the selected sample (N = 29) 21 papers published in academic journals demonstrate the average impact factor 2 with *Research Policy* having the highest impact factor (5.68). The remaining eight papers published in conferences' proceedings. e.g. Knowledge Management International Conference (Kmice) 2016.

Table 1. Publications over time.

Journal	No. of articles	Impact Factor
Technological Forecasting and Social Change	3	2.678
International Journal of Technology Management	2	1.19
International Journal of Arts Management	2	0.69

4.1.3 Papers by Methodology

Figure 3 presents research methods used in the selected sample. As it can be seen from the chart 37% of papers (11) gathered data through survey/interviews techniques; 33% (10) of papers applied such techniques as data modelling, critical discourse analysis and netnography; 30% (8) of papers applied case-study analysis.

Even though the spread of techniques used is relatively even it can be inferred that the research area is recent and still needs to be well established and conceptualised as currently most of the data is obtained through interviews and surveys.

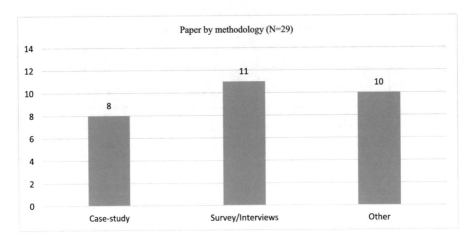

Fig. 3. Paper distribution by methodology.

4.1.4 Papers by Industry Type

Figure 4 represents allocation of the selected sample by industry. As mentioned earlier creative industries represent a broad range of activities from technological to non-technological. For the analysis to be accurate this study adopted internationally recognised classification by DCMS which is updated annually.

As it can be seen from the chart below the highest number of papers (6) that analyse open innovation technologies in creative industries were studied in the context of publishing, e.g. book sector, newspaper. This can be explained by two reasons. Firstly, publishing industry has undergone major changes since the proliferation of IT technologies. For example, the role of Twitter in the Arab uprisings. Secondly, publishing sector represents content producing sector and hence it is easier to measure impact of IT and adoption of innovative technologies such as open innovation. The second

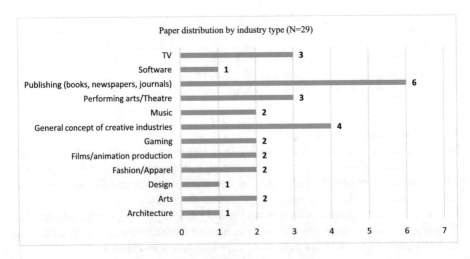

Fig. 4. Paper distribution by industry type.

largest group of papers (4) was analysed in a general context of creative industries or few creative sectors combined followed by papers (3) in a context of performing arts and papers (3) in a context of TV.

4.2 Content Analysis

In order to provide a better understanding of the current research area this section presents a content analysis of the 29 papers. Following the main objective to what extent do creative industries adopt open innovation technologies the following research questions were identified: (1) How do creative industries adopt open innovation technologies??; (2) What are factors that influence the adoption of open innovation technologies in creative industries?; (3) How do adopted open innovation technologies impact creative industries?

The findings are introduced below in the following subsections.
How do creative industries adopt open innovation technologies?

Twelve papers were identified in order to understand how creative industries adopt open innovation technologies and what kind of technologies are mostly used. Bruns and Humphreys [27], Thoren et al. (2014) [28], Mihaljevic (2015) [29], Bulock and Watkinson (2017) [30] analyse application of open innovation technologies in publishing and content creation sector. Bruns and Humphreys (2010) [27] analyse an open and collaborative digital platform where users can upload and share media content to develop themselves into innovative creators. The application of this platform was analysed in relation to economic development of a particular community in Australia. Thoren et al. (2014) [28] studied open practices in newspaper industry in Sweden where open innovation is supported primarily by social media. Study reveals that in newspaper sector editors consider users only as content providers and gatekeeping policy is still widely practices in the sector. Mihaljevic (2015) [29] and Bulock and Watkinson (2017) [30] research how open innovation technologies shape book industry, e.g. crowdfunding to facilitate open access in digital libraries. Research calls for more collaboration between universities and digital libraries.

Boeuf et al. (2014) [31], Faludi (2015) [32], Mollick and Nanda (2016) [33] discuss open innovation technologies in the area of performing arts, e.g. theatre. Boeuf et al. (2014) [31], Mollick and Nanda (2016) [33] show that in case of theatres and independent plays organisations are more likely to use crowdfunding platforms, e.g. Kickstarter. Results demonstrate that actors involved in crowdfunding projects see it as a non-commercial activity and crowdfunded projects demonstrate higher chances to be realised than the ones suggested by experts. On other hand Faludi (2015) [32] argues that in performing arts stylistic innovations are more important than technological and open innovation practices are primarily adopted in a form of external knowledge, e.g. workshops while technology is seen as an enabler for knowledge.

Schuurman et al. (2011) [34] and Fortunato et al. (2017) [35] research open innovation technologies in TV. Schuurman et al. (2011) [34] conduct a case study of 'Living-lab' concept introduction for mobile TV. Firstly, authors suggest the link between open innovation and 'Living-lab' technologies. Secondly, authors believe that application of open innovation technologies may help TV companies to tackle problems associated with ICT challenges. Fortunato et al. (2017) [35] propose application of big data techniques to improve open innovation practices for social TV where social media is used as open innovation technology.

Haefliger et al. (2010) [36] explore open innovation technologies in a form of collaborative platform in gaming industry where users are invited to create their own content for motion production. Authors state that collaborative technologies shape current gaming industry as before such software was too expensive and was only available for professionals while now technology has lowered entry barriers for new users.

The last two papers in this sample deal with open source software and music. Joo et al. (2012) [37] analyse application of open source software as a part of open innovation concept while Zabaljáuregui and Dorado (2016) [38] run an experiment for collaborative music creation via Opensemble-platform.

What are factors that influence the adoption of open innovation technologies in creative industries?

In relation to the second research question five papers were selected that deal with factors that might positively or negatively impact open innovation practices in creative sectors. Stejskal and Hajek (2016) [39] identify that in a context of creative industries innovations are more effectively produced when companies collaborate with external parties. Moreover, creative industries create knowledge spillover through external collaborations. Hobbs et al. (2016) [40] identify network size and project quality as key predictors that positively impacts the success of crowdfunding platforms in independent film industry. Another study by Bao and Huang (2017) [41] reveal the following factors that impact positive adoption of open innovation technologies. These factors are reward support, impression support, and relationship support that facilitate positive outcomes of crowdfunding in creative industries. In the case with architecture Lazzeretti et al. (2016) [42] suggest that involving local artists' when introducing new laser technology in Florence has higher positive outcomes than when solely relying on R&D. This is explained by the factor that local artists and cultural workers are keener to share knowledge and experience as they have strong cultural attachment to the city. Among negative factors that lead to failure of opening up R&D practices in creative industries Benghozi and Salvador (2016) [43] bring lack of investments into new information technologies and application of old business models in publishing sector.

How do adopted open innovation technologies impact creative industries?

The last subsection of the content analysis includes 12 papers that analyse the impact of adopted open innovation technologies within creative industries. Hracs (2012) [44] analyses impact of information technologies and digital platforms in music industry. This impact is seen as twofold. From one side emerging information technologies facilitated the rise of independent musicians. From other side competition among young musicians has also increased as anyone can collaborate towards music creation process. Another creative industry that has undergone changes with fast development of information technologies is fashion. Two studies highlight positive impact of open collaborative platforms. Pihl and Sandstrom (2013) [45] explore openness in business development through fashion bloggers in Sweden. Results obtained through study demonstrated that brands that cooperate with bloggers reduce transaction costs while increase ROI. Furthermore, Scuotto et al. (2017) [46] analyse positive impact of open innovation technologies on ROI in smart fashion industry. However results suggest that companies lack understanding on how to manage digital technologies in the context of creative industries.

Le et al. (2013) [47] explore the link between technology and creative process in gaming industry (Ubisoft Open World Platform) on micro level. Authors argue in creative industries impact of technologies is mostly studied in a context of data collection and for marketing purpose and little attention is paid on impact on creative product itself. Holm et al. (2013) [48] investigate degree of openness and its impact on newspaper sector in Denmark. The results reveal a decline happening in the sector through transformation of current business models towards becoming more open, dynamic and multichannel. Mangematin et al. (2014) [49] introduce multidimensional framework that demonstrates positive impact of digital platforms on content-producing industries, e.g. music, publishing. Sung (2015) [49] compare impact of information technologies in manufacturing and creative industries. Results highlight that unlike in manufacturing industries in creative industries better understanding of IT efficiency and threats it poses is needed.

Interesting to note that two studies in the selected sample adopt service science perspective to analyse impact of open innovation technologies on value creation process in creative industries. Novani et al. (2015) [50] research how open collaboration in batik design can positively impact solo tourism while Quero et al. (2017) [51] explore impact of crowdfunding in arts on shaping value-in-context by deriving 7 types of value. On other hand Brabham (2017) [52] argues that mass 'democratisation' of digital platforms like crowdfunding poses threat to arts when it comes for obtaining funds from governments.

Lin (2015) [53] explores positive impact on TV production through opening up backstage data to public through open innovation technologies in case with BBC. Study demonstrates that such approach contributes towards emergency of 'techno-elite' audience. Finally, Loriguillo-Lopez (2017) [54] infers that open innovation technologies and platforms promote globalisation of cultural goods. For example, crowdfunding has raised global interest towards anime in Japan as more fans feel emotionally connected to products they fund.

In summary the results obtained through systematic review of 29 papers revealed the following. Firstly, descriptive analysis demonstrated the growing interest towards research area among scientific community in recent years. Latter point can be also supported by the fact that the highest number of papers (21) are published in journals rather than in proceedings. Furthermore, research techniques used in the selected sample show a need for more empirical research. On other hand content analysis showed that the most studied technology associated with open innovation is crowdfunding. This point can be explained by the fact that some scholars consider crowdfunding as a facilitator for inbound open innovation processes (Stanko and Henard, 2017) [55]. In addition to that findings showcased that despite technological transformation of some creative sectors companies do not understand how to correctly fit and open up business models towards becoming more consumer-centric. Misunderstanding and misuse of digital technologies might be threating to traditional business, e.g. newspaper. Finally, open innovation and information technologies in general not only pose a disruptive change to such industries like music, but also contribute to globalisation of very specific cultural goods like anime.

Based the above scenario we identified the two directions for future research in the area. Firstly, the descriptive analysis showed that majority of studies (6) in the selected sample are focused on publishing sector. However it would be interesting to explore the concept of open innovation in fashion context. For example, how do large and well-established brands embrace open innovation technologies? Secondly, prior research on open innovation is mostly focused on analysing inbound open innovation activities, i.e. bringing external knowledge or technology in-house while neglecting outbound open practices. However creative industries represent interesting research domain to explore how do technologies and innovations in a form of intellectual properties produced by creative industries applied for outbound open exploitation processes?

5   Conclusion

This study presented a systematic review of literature on adoption of open innovation technologies in the context of creative industries.

The first section provided a literature overview on open innovation, open innovation technologies and creative industries along with defining current research gaps and explaining the importance of the research area. The second section presented an overview of methodology and selection process of 29 papers in line with the guidelines outlined [25]. The third section delivered two-stage report based on the results obtained from 29 papers in a form of descriptive and context findings. The current analysis determined that the research area poses an interest to scientific community, but it is still in the early stages. Application of open innovation technologies in creative industries is limited to certain industries where impact is rather obvious and innovation process is linear.

From theoretical perspective this paper addressed the current research gaps in the field of open innovation and expanded its research domain onto creative industries. Thus, it can serve as a starting point for future research in the area. From practical perspective this study provided a guidance for creative managers on how open innovation technologies should be approached in creative industries and impact of cultural bond in certain areas. For example, working closely with local artists and community in preserving cultural heritage of cities than relying solely on R&D.

Finally, to the best of authors' knowledge this paper is one of the first to systematically analyse application of open innovation technologies in different sectors of creative industries, but it has certain limitations. Firstly, future work in the area can be focused entirely on a particular sector within creative industries, e.g. impact of open innovation on performing arts. Secondly, for richer overview papers can be extracted from top journals in innovation management field rather than one database.

Acknowledgments. We gratefully acknowledge Research Grants Council and The Hong Kong Polytechnic University under a project account code RU6S for supporting this research.

References

1. Van de Vrande, V., de Jong, J.P.J., Vanhaverbeke, W., de Rochemont, M.: Open innovation in SMEs: trends, motives and management challenges. Technovation **29**, 423–437 (2009)
2. Enkel, E., Gassmann, O., Chesbrough, H.: Open R&D and open innovation: exploring the phenomenon. In: Enkel, E., Gassmann, O., Chesbrough, H. (eds.) R&D Management. Special Issue: Open R&D and Open Innovation, vol. 39(4), pp. 311–316 (2009)
3. The International Confederation of Societies of Authors and Composers (CISAC). http://www.worldcreative.org/#overview
4. UNCTAD: Development Report 2008, New York and Geneva, pp. 31–40 (2008)
5. Department for Culture, Media and Sports (DCMS): Creative Industries Economic Estimates. DCMS, London (2016). https://www.gov.uk/government/uploads/system/uploads/attachment_data/file/523024/Creative_Industries_Economic_Estimates_January_2016_Updated_201605.pdf
6. Chesbrough, H.W.: Open Innovation: The New Imperative for Creating and Profiting from Technology. Harvard Business School Press, Boston (2003)
7. Bogers, M., Horst, W.: Collaborative prototyping: Crossfertilization of knowledge in prototype-driven problem solving. J. Prod. Innov. Manag. **31**(4), 744–764 (2014)
8. Gassmann, O., Enkel, E.: Towards a theory of open innovation: three core process archetypes. In: Proceedings of the R&D Management Conference, Lisbon, Portugal, 6–9 July 2004 (2004)
9. Lichtenthaler, U., Lichtenthaler, E.: A capability-based framework for open innovation: complementing absorptive capacity. J. Manage. Stud. **46**(8), 1315–1338 (2009)
10. Hrastinski, S., Ozan, H., Kviselious, N., Edenius, M.: A review of technologies for open innovation: characteristics and future trends. In: Proceedings of the 43rd Hawaii International Conference on System Sciences, Poipu, HI, 5–8 January 2010 (2010)
11. Gordon, S., Tarafdar, M., Cook, R., Maksimoski, R., Rogowitz, B.: Improving the Front End of Innovation with Information Technology. Res. Technol. Manage. **51**(3), 50–58 (2008)
12. Lundström, J.S.E., Wiberg, M., Hrastinski, S., Edenius, M., Agerfalk, P. (eds.): Managing Open Innovation Technologies. Springer Science & Business Media, Heidelberg (2014)
13. Stanko, M.A, Fisher, J.G., Bogers, M.: Under the Wide Umbrella of Open Innovation. Journal of Product Innovation (2017)
14. Serrat, O. Harnessing creativity and innovation in the workplace. In Knowledge Solutions; Asian Development Bank: Manila, Philippines, pp. 61–72 (2017)
15. Stierand, M. and Dörfler, V.: Methods against methods. Technology for Creativity and Innovation: Tools, Techniques and Applications, pp. 121–134 (2011)
16. Stierand, M., Dörfler, V., Macbryde, J.: Creativity and innovation in haute cuisine: Towards a systemic model. Creativity Innov. Manage. **23**(1), 15–28 (2014)
17. McAfee, A., Dessain, V. and Sjoman, A., Zara: IT for fast fashion. Harvard Business School, case # 9-604-081, (2004)
18. Béraud, P., Du Castel, V., Cormerais, F.: Open innovation, economy of contribution and the territorial dynamics of creative industries. J. Innov. Econ. Manage. **2**, 81–105 (2012)
19. Young, V.M.: Bacelar's TheCurrent Pushes Open Innovation (2017). http://wwd.com/fashion-news/fashion-scoops/liz-bacelar-thecurrent-pushes-open-innovation-h-farm-fashion-technology-accelerator-sxsw-fashion-scoop-10840019
20. Lichtenthaler, U.: External technology commercialisation projects: objectives, processes and a typology. Technol. Anal. Strateg. Manag. **20**(4), 483–501 (2008)
21. Parida, V., Westerberg, M., Frishammar, J.: Inbound Open innovation activities in high-tech SMEs: the impact on innovation performance. J. Small Bus. Manage. **50**(2), 283–309 (2012)

22. Garnham, N.: From cultural to creative industries: an analysis of the implications of the "creative industries" approach to arts and media policy making in the United Kingdom. Int. J. Cultural Policy **11**(1), 15–29 (2005)
23. Hesmondhalgh, D.: The Cultural Industries, 2nd edn. Sage, Thousand Oaks (2007)
24. Potts, J., Cunningham, S.: Four models of the creative industries. Revue d'économie politique **120**(1), 163–180 (2010)
25. Tranfield, D., Denyer, D., Smart, P.: Towards a methodology for developing evidence-informed management knowledge by means of systematic review. Br. J. Manag. **14**(3), 207–222 (2003)
26. Howkins, J.: The Creative Economy: How People Make Money From Ideas, Penguin (2001)
27. Bruns, A., Humphreys, S.: Research adventures in Web 2.0: encouraging collaborative local content creation through Edgex, Media Int. Aust. (136), 42–59 (2010)
28. Thorén, C., Ågerfalk, P.J., Edenius, M.: Through the printing press: an account of open practices in the swedish newspaper industry. J. Assoc. Inform. Syst. **15**(11), 779 (2014)
29. Mihaljević, J.: An Academic Library Model in the Creative Industries System of the Digital Age. Medunarodni Znanstveni Simpozij Gospodarstvo Istocne Hrvatske - Vizija I Razvoj (2015)
30. Bulock, C., Watkinson, C.: Opening the Book. Serials Review, vol. 43, issue 2 (2017)
31. Boeuf, B., Darveau, J., Legoux, R.: Financing creativity: crowdfunding as a new approach for theatre projects. Int. J. Arts Manage. **16**, 33–48 (2014)
32. Faludi, J.: Open innovation in the performing arts. Examples from contemporary dance and theatre production. Corvinus J. Sociol. Soc. Policy **6**(1), 47–70 (2015)
33. Mollick, E., Nanda, R.: Wisdom or madness? Comparing crowds with expert evaluation in funding the arts. Manage. Sci. **62**(6), 1533–1553 (2016)
34. Schuurman, D., Moor, K.D., Marez, L.D., Evens, T.: A Living Lab research approach for mobile TV. Telematics Inform. **28**(4), 271–282 (2011)
35. Fortunato, A., Gorgoglione, M., Petruzzelli M.A., Panniello, U.: Leveraging big data for sustaining open innovation: the case of social TV. Inform. Syst. Manage. **34**(3), 238–249 (2017)
36. Haefliger, S., Jager, P., von Krogh, G.: Under the radar: industry entry by user entrepreneurs. Research Policy **39**, 1198–1213 (2010)
37. Joo, C.H., Kang, H., Lee, H.: Anatomy of open source software projects: evolving dynamics of innovation landscape in open source software ecology. In: The 5th International Conference on Communications, Computers and Applications (MIC-CCA2012), Istanbul, Turkey (2012)
38. Dorado, D., Zabaljauregui, M.: Opensemble: a framework for collaborative algorithmic music (2016)
39. Stejskal, J., Hajek, P.: Determinants of collaboration and innovation in creative industries a case of the Czech Republic. In: Proceedings of Knowledge Management International Twenty 29–30 August 2016, Chiang Mai, Thailand (2016)
40. Hobs, J., Grigore, G., Molesworth, M.: Success in the management of crowdfunding projects in the creative industries. Internet Res. **26**(1), 146–166 (2016)
41. Bao, Z., Huang, T.: External supports in reward-based crowdfunding campaigns: a comparative study focused on cultural and creative projects. Online Inform. Rev. **41**(5), 626–642 (2017)
42. Lazzeretti, L., Capone, F., Cinti, T.: Open innovation in city of art: the case of laser technologies for conservation in Florence, City, Culture and Society, vol. 2, pp. 159–168 (2011)

43. Benghozi, P.J., Salvador, E.: How and where the R&D takes place in creative industries? Digital investment strategies of the book publishing sector. Technol. Anal. Strateg. Manage. **28**(5), 568–582 (2016)
44. Hracs, B.J.: A creative industry in transition: the rise of digitally driven independent music production. Growth Change **43**, 442–461 (2012)
45. Pihl, C., Sandstrom, C.: Value creation and appropriation in social media - The case of fashion bloggers in Sweden BT. Int. J. Technol. Manage. **61**(3–4), 309–323 (2013). Special Issue on Opening Up Innovation and Business Development Activities
46. Scuotto, V., Giudice, M.D., Della Peruta, M.R., Tarba, S.: The performance implications of leveraging internal innovation through social media networks: an empirical verification of the smart fashion industry. Technol. Forecast. Soc. Chang. **120**, 184–194 (2017)
47. Lê, P., Massé, D., Paris, T.: Technological change at the heart of the creative process: insights from the videogame industry. Int. J. Arts Manage. **15**(2), 45–86 (2013)
48. Holm, A.B., Günzel, F., Ulhøi, J.P.: Openness in innovation and business models: lessons from the newspaper industry. Int. J. Technol. Manage. **61**(3–4), 324–348 (2013)
49. Sung, T.K.: Application of information technology in creative economy: manufacturing vs. creative industries. Technol. Forecast. Soc. Chang. **96**, 111–120 (2015)
50. Novani, S., Putro, U.S., Hermawan, P.: An application of soft system methodology in batik industrial cluster solo by using service system science perspective. Procedia-Soc. Behav. Sci. **115**(1), 324–331 (2014)
51. Quero, M.J., Kelleher, C., Ventura, R.: Value-in-context in crowdfunding ecosystems: how context frames value co-creation. Serv. Bus. **11**(2), 405–425 (2017)
52. Brabham, D.C.: How crowdfunding discourse threatens public arts. New Media Soc. **19**, 1–17 (2016). https://doi.org/10.1177/1461444815625946
53. Lin, Y.: Open data and co-production of public value of BBC backstage. Int. J. Digital Telev. **6**(2), 145–162 (2015)
54. Loriguillo-López, A.: Crowdfunding Japanese commercial animation: collective financing experiences in anime. Int. J. Media Manage. **19**(2), 182–195 (2017)
55. Stanko, M.A., Henard, D.H.: Toward a better understanding of crowdfunding, openness and the consequences for innovation. Res. Policy **46**, 784–798 (2017)

Information Technology in Management Science

The Concept of Management Accounting Based on the Information Technologies Application

Nadezhda Rozhkova[1]([✉]), Uliana Blinova[2], and Darya Rozhkova[2]

[1] State University of Management,
Ryazanski prospect, 99, Moscow 109542, Russia
nakoro@yandex.ru
[2] Financial University under the Government of the Russian Federation,
Leningradski prospect, 49, Moscow 125993, Russia
rosbu@mail.ru, rodasha@mail.ru

Abstract. Based on the analysis of foreign and domestic experience, main objectives of the organization of management accounting system in modern business have been discussed, the need of creation an information system of management accounting has been proved. A historical review of development methods of automating management accounting has been carried out since 1950s to the present. It is revealed that modern information technologies in the field of automation of management accounting are represented by two large groups - local and integrated systems. It is noted that local systems should be used in small business, and integrated ones in medium and large businesses. An expediency of using integrated ERP-class management automation systems is proved.

Keywords: Enterprise economics · Management accounting
Information systems · Enterprise resource planning system (ERP)
Automated control system (ACS)

1 Introduction

Recently, serious technological changes, including the emergence of Internet technologies, bring unique characteristics to global economic system, as well as to entrepreneurial activity.

Management accounting is a universal information system that meets diverse information needs of enterprise management. To form efficient accounting procedures in the field of consumption of funds in economic activities, enterprises need to have a regulated subsystem of management accounting. In this subsystem, information technology has a paramount importance.

The study of the literature showed the absence of systematic research on the organization of management accounting in the field of information technology.

On the one hand, insufficient knowledge and degree of elaboration of the organization of management accounting in the field of information technologies, on the other hand, scientific and practical importance, have determined the choice of topic of this

© Springer International Publishing AG, part of Springer Nature 2018
T. Antipova and Á. Rocha (Eds.): MosITS 2017, AISC 724, pp. 89–95, 2018.
https://doi.org/10.1007/978-3-319-74980-8_8

study, its purpose, object and subject, as well as the range of issues under consideration.

The relevance of the study is due to the fact that digital economy is by nature not static, it evolves, new phenomena are emerging that affect all sectors and types of economic entities, as well as the nature of the production and financial activities of market actors.

2   Main Part

In modern market conditions the system of management accounting acts as an information basis for management, since it is the basis for taking effective management decisions. Managers of enterprises make significant efforts to establish and maintain management accounting, but it is required an interest of managers and specialists of enterprises, as well as organizational prerequisites and conditions for the management accounting implementing.

When analyzing foreign and domestic scientific researches [1–4], we identified three main objectives of the organization of the management accounting system in modern business (Fig. 1).

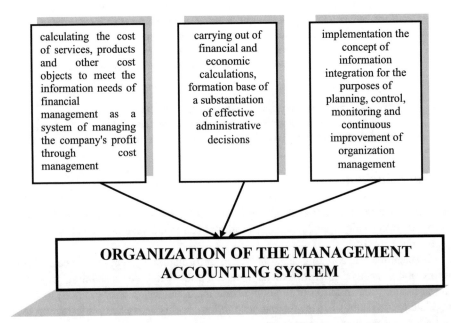

Fig. 1. The main objectives of the organization of the management accounting system.

We want to underline that if a certain level of managerial culture is formed in an enterprise, the need to organize an information system of management accounting is increasing. Simultaneously, managers of organizations face with existing management system limitations. To begin with, it is lack of an effective enterprise planning mechanism that can allow to carry analysis of future decisions to be made, to calculate economically justifiable cost indicators, and to analyze variances of actual indicators from planned ones, to identify the reasons of deviations.

Second problem is an absence of a reliable and objective system of cost accounting, which will enable to analyze the costs in different areas, and, ultimately, effectively control the activities of the company and its management. Third limitation is an inefficient internal reporting system that does not adequately reflect the requirements of management tasks.

A significant drawback is the lack of a mechanism for assessing the profitability of an enterprise in general and each type of product and service in particular. It is possible to consider also the human factor, such as insufficient level of responsibility and motivation of personnel to minimize company costs.

To eliminate the identified limitations, researchers representing different scientific schools offer numerous directions from the development of forms of primary documents and internal reporting to the methodology of management accounting [5]. However, any document and any technique requires certain efforts from the human factor's point of view. And, most importantly, a complexity of the management accounting system should be reduced and not vice versa.

By virtue of scientific and technical progress, a process of "digitalization" is taking place, which includes storage of data in digital form. Unlike the economy of the past years, when the information was analogue (physical), relationships of market participants were only possible through their actual communication. In modern realities with help of digital devices there is a free movement of information in unlimited quantity and in shortest possible time between people in different parts of the world.

Information technologies directly affect utilization of management accounting system. Modern information technologies of accounting and enterprise management, as an instrument of management accounting, allow to implement many methods of management accounting, create a single information space in the organization and thereby provide both management and accounting with information that meets the high demand of modern business organization.

From the perspective of integrating heterogeneous information which is formed in the enterprise for management, it is necessary to adhere classical principles and characteristics, information requirements, among which, as a rule, are distinguished: a form of information submission - information must be understood by a specific user; frequency - the submission of information should be delivered regularly; accuracy of information - an acceptable compromise between reliability of data and timeliness of presentation; a clear definition of a particular employee responsibility to prepare management information in required form, with a certain accuracy (reliability) and its transfer at the due time to the recipient; efficiency or profitability - benefits of the entire chain of management reporting (data - summaries - reports - summary report) should exceed the costs of its implementation.

While building complex management accounting systems covering all levels of management, these requirements, undoubtedly, dictate the need of automation of accounting procedures, since manual processing of data does not allow to provide the required quality of information.

More than that, management accounting has always been oriented towards efficiency, which can only be achieved through the use of information technology and automation.

Scientists and practitioners have been thinking about the automation of management accounting since the 50-ies of the last century. Manual processing of data in the conditions of numerous product groups, distribution of standard costing method caused the need for new software algorithms, automation of managerial accounting.

However, technological possibilities to a large extend has limited possibilities of automation. Further development of complex computer installations, personal computers in the 80-ies of the last century allowed domestic scientists, in particular representatives of Kazan school of management accounting, K.M. Garifullin, V.B. Ivashkevich, and others to develop the concept of an automated control system (ACS) [6].

The essence of the ACS concept is an integration of management information of the enterprise on the basis of unified algorithms and databases.

Crisis phenomena in the economy of Russia in the 90s did not significantly restrain the development of the practice of management accounting and its information and analytical aspects. Nevertheless, in case of small volumes of business operations, management accounting was still conducted using simple Excel tables.

As the enterprise developed, the number of business transactions increased, the amount of financial resources investing in information technology and software has changed accordingly. As a result both foreign (Axapta, Exact, Platinum, Hansa Solutions, Scala, Accpac, SunSystems) and Russian systems of management accounting automation (BOSS Corporation, Galaxy, Parus, Magnat, Alfa, Etalon, Inotec, 1C) has appeared and been implemented in day-to day activity of enterprise.

Also in the early 90s, efforts of the consulting company "Gartner Group" resulted in a term "ERP-system", which refers to a enterprise resource planning information system designed to automate the accounting and management of all resources, costs and results of enterprise.

The ERP-system can be considered as an information generation system for management purposes, an information basis for effective management of all resources of organization. As a rule, ERP systems are built on a modular basis and to some extent cover all key processes of company's activities.

At the same time, development of information technologies in the field of automation of management accounting, in general took place as creating local and integrated systems. In local systems, only certain accounting functions have been automated, such as in the Russian program "Info-bughalter". In systems of this class it was not possible to implement functions of management accounting in full, since they contain limited functions for accounting of analytical information, and the interrelations between organizational units are not realized. In general, local systems are convenient for small businesses, but not acceptable for medium and large companies with extended requirements for management information.

Some systems, such as "1C" products, began with the stand-alone solution "Accounting" and, in the future, began to represent an integrated solution based on "1C: Accounting 8" programs and individual configured and functionally expanded subsystems ("1C: Enterprise 8. Management of corporate finance", "1C: Enterprise 8.

Management of manufacturing enterprise"). This allowed for a relatively effective management of investments of both enterprises and groups of companies (holdings, conglomerates), control expenditures and, significant improvement of manageability an entire business and its competitiveness.

At present, integrated systems are widely used, they differ in terms of functionality and cost.

It is most effective with help of ERP-class management automation systems to create a single data warehouse, containing all necessary information about the enterprise: services provided, products produced, costs and results of each area of activity, resources consumed at all stages of supply, production and marketing, an amount of compensation paid to different categories workers and employees, an information about all enterprise business-unites, etc.

The main functions of management accounting implemented in various modules of ERP-systems include the implementation of management accounting procedures for costs, results, and resources of the enterprise used at all stages, including procurement and procurement activities, production, sales, advertising, investment, auxiliary production, social services, etc. In particular:

- accounting of material resources and production: maintenance of design and technological specifications that determine composition of manufactured products, including calculations of volume of material resources and duration (labor intensity) of operations necessary for their production; formation of resource requirements, budgeting of sales and production; analysis of materials and components need, terms and volumes of supplies to implement a production plan; inventory management and procurement; contract management, centralized procurement, accounting and optimization of warehouse resources;
- accounting of production capacities from large-scale budgeting to use of individual machines and equipment;
- providing information in order to operational management of finances, including the preparation of a financial plan and monitoring its execution,
- management of investment projects, including assessment of the profitability of enterprise on the basis of flows of financial, labor, material and other resources analysis.

The software tools used in ERP-systems allow applying modern methods of management accounting, structuring a flow of orders, assessing possibility of their implementation in departments and divisions, linking it with sales, creating a database for managing costs, results, resources and profitability of enterprise as a whole.

By integrating separate subsystems of the software product, a broad base of information support for the management process is formed, which aims to optimize result while maintaining the necessary liquidity. An end-to-end information flow is provided between individual objects and processes, as well as between operational and strategic information subsystems. The top management of the enterprise always has an idea about processes taking place in the divisions, even in conditions of decentralization of management.

Using the concept of management accounting using ERP-systems is especially relevant for organizations that:

- has a lack of clear mechanism to manage business units: a system of key indicators, regulations for their planning; reporting, analysis and evaluation, motivation;
- there is no effective mechanism of planning activities, allowing preliminary comparative analysis of decisions, calculating planned, economically justified (in accordance with internal norms and standards) cost indicators, forecasting results of operations and justifying long-term solutions, analyzing deviations of actual indicators from planned ones and identifying their causes;
- absence of procedures to conduct analysis and make managerial decisions related to formation of production program, pricing, evaluation of investment projects, etc.
- lack of transparent cost accounting system that does not allow to determine both the reliable value of costs and analyze them by types, places of origin, responsibility centres and in other sections necessary for adequate control of activities and management;
- imperfection of internal reporting system;
- lack of a mechanism for assessing the profitability of business and individual products;
- inadequate level of responsibility and motivation of personnel to reduce level of costs and improving efficiency of activities of both their own unit and an enterprise as a whole.

At the same time, an implementation of ERP-system is quite an expensive process, which requires a balanced and thoughtful solution. It should be a tool for decision-making, and not an additional burden on all units of organization.

At the beginning of application management accounting concept based on the use of ERP-system, first of all it is necessary to describe all business processes of the company, reorganize its activities in accordance with business logic embedded in the system, reengineering business processes. It is not just implementing a relevant program, but rather, a complete reorganization of entire internal environment of management accounting.

Therefore, some measures are required to change core activities aimed for more complete compliance with a logic embedded in a system, which will optimize a scheme of functions and operations performed by employees of organization in relation to automated workstations created in new configuration of system.

Also, to increase an effectiveness of management accounting using the ERP-system, it is necessary to integrate all economic management functions at all levels of management; parallel automation not only management accounting, but also financial and tax accounting.

3 Conclusion

Usage of modern computer technologies for processing economic information, a comprehensive presentation of content of management accounting is not only possible, but also the most optimal. Development of management accounting will largely depend on evolution of computer technology, but, of course, the main were, and will be the needs of management, composition of its tasks and emerging problems.

In general, an application of concept of management accounting based on information systems allows to achieve main objectives of management accounting creation in today's business: cost calculating of services, products in order to meet an information needs of financial management as a system of managing the company's profit through cost management; carrying out financial and economic calculations, base formation of substantiation of effective administrative decisions; implementation of information integration for the purposes of planning, control, monitoring and continuous development of organization management.

References

1. Yu, B.U., Rozhkova, N.K., Yu, R.D.: Evaluation of Entrepreneurial Activity and Management Accounting. Financial University (2010). 134 p
2. Azirilian, A.N., Azriliyan, O.M., Kalashnikova, E.V., Kvadrakova, O.V.: Big Economic Dictionary, 7 edn., supplemented. Institute for New Economy, Moscow (2014). 1472 p
3. Mizikovsky, I.E.: Management accounting. In: Mizikovskiy, I.E., Miloserdova, A.N., Frolova, E.B., Mizikovsky, E.A. (ed.) Educational - Methodical Manual, 101 p. Nizhny Novgorod State University, Nizhny Novgorod (2008)
4. Yu, R.D.: Developing of the system of management accounting of entrepreneurial activity in international tourism, vol. 10, pp. 210–214. State University, State University of Management (2016)
5. Kasurinen, T.: Exploring management accounting change: the case of balanced scorecard implementation. Management accounting research. – 2002. – T. 13. – NO. 3. – C. 323–343
6. Garifullin, K.M., Ivashkevich, V.B.: Accounting Financial Records: Textbook. Publishing House KFEI, Kazan (2002)

Management of Financial Bubbles as Control Technology of Digital Economy

Gennady Ross$^{(\boxtimes)}$ and Vladimir Liechtenstein

Financial University, Moscow, Russia
ross-49@mail.ru

Abstract. Discusses the growing role of financial bubbles in the digital economy, which is characterized by the use of artificial intelligence (AI) and the Internet for decision-making on the basis of raw data. It is shown that the classical market paradigm, according to which supply and demand are balanced through price, does not work. Proposed a new paradigm based on the theory of equilibrium stochastic processes (ESP), which argues that economic agents are guided by their risks. The classical paradigm is a special case of the new paradigm. Under the new paradigm offers ways of measuring financial bubbles and management by it.

Keywords: Financial bubbles · Digital economy · Management
Market · Equilibrium paradigm · The evolutionary-simulation methodology
The equilibrium stochastic process

1 Introduction

Any activity is easily converted into inflating of a bubble. There are innumerable methods of such a transformation: the organization of the rush demand, brand promotion, monopolization, and theft, distortion, the creation of the pyramids, financial terrorism, etc. The essence of a financial bubble is that the price of the goods, services or securities greater than that which would have to be if they meet the conditions of perfect competition. The difference between these prices and there is a bubble. In other words, the bubble is assets not, which secured. Any way of making money from money, e.g. receiving interest on loans, is, generally speaking, bloating financial bladder and prohibits Islamic banks (https://www.syl.ru/article/332940/islamskiy-bank-printsip-rabotyi-osnovnyie-pravila-islamskiy-bank-bez-protsentov-bankovskaya-deyatelnost).

Based on the theory of financial bubbles Fry, J., & Brint, A. develop a model for over-confidence in opinion polls and betting odds. Here, over-confidence manifests itself in the sense that, collectively, market participants are over-precise in their assessments of the underlying price risk. Their approach copes well with known features of polling data such as irregularly-spaced time series. Further, the elegance of our approach is such we can easily make adjustments for the size of the distortion found. [2]

Financial bubble is a useful, desirable and even required for the health of the economy if it is not too big and not too stable. It is easy to explain with a simple example, which was discussed in one of the news programs: the girl invented the folding Slippers. She took out a loan, have established small-scale production and the product itself was advertised through "word of mouth". Thus arose the business that has created more than 100 jobs. This example shows that excessive demand may occur spontaneously, that he may establish useful business and that without the loans would not have happened. In General, any attempt to inflate the bubble is motivated by the desire to earn easy money is a inescapable impetus for economic activity. If you imagine a kind of ideal economy in which managed to completely eradicate the bubbles, it will be either stagnating or deteriorating economy.

Problems create not the bubbles themselves, and even their swelling. In a healthy economy bubbles deflate by themselves. Usually this is due to the recession hype. The problems created by activities aimed at preserving bubbles. This is of interest not only enthusiasts of easy money, but also a business that is already in some part of it turned into a bubble, and its numerous lobbyists in the legislative and Executive power. For achieving to goal to keep the bubble can serve so financial strategy as refinancing, rebranding, marketing innovation, issue of securities etc. Bubbles can break and merge, to mimic, to adhere to the actual production and break away from him. If the bubble is sufficiently resistant to changes in the external environment, market fluctuations and financial stress, it begins to play the role of a safe haven, raise capital, absorb the actual production.

2   Bubble Management System

The main negative impact of the financial bubble is:

- first, the bubble displaces the price into right, that is, introduces an error in the information of prices, distorts economic measurement and distorts the calculations;
- second, a creating unwanted capital flows, the concentration of capital in the bubbles, distortions in the economy and the stratification by income;
- third, creating uncertainty for economic agents (EA), increases the risks for investments and transactions, leads to chaos in the economy.

In the digital economy role of financial bubbles is multiplied due to the fact that it is based on the massive use of artificial intelligence and the Internet for decision-making by processing the raw data. German Gref believes that "Data becomes information as the results of analyze. New technologies allow to process raw data ... the maelstrom Whirlpool will swirl all data, and even human nature. ... 230 likes and we will understand your personality better than friends. It is used to manage client relations. Already, you can read psycho from the pictures ... It's a fundamentally new model of identifying true human needs."[1] Today, the trading companis, for example, "Ali Baba" have ways to in real time to identify preferences and to predict the wishes of the buyers. It wasn't on the previous technological structures, when created machines that can move faster than people, to work more productive than a man to do what man is not able, for example, to fly. Now the machines have learned to think better than person.

---

[1] https://www.youtube.com/watch?v=mcs_GFgthhs

It became apparent that artificial intelligence has surpasse human intelligence in speed of information processing, memory, and functionality when working with raw data. Another thing is that "a Machine that generates text, or an image, does not understand that it generates. It generates a sequence of bytes. We have already filled this sequence semantical, determining that this image looks like a picture of Rembrandt..."[2] «Intelligent systems have the ability to ... training, rational distribution and directed or intelligent behavior. These 2 components they have, there is 1st: they can to learn. We can't simulate human intelligence fully, whoever he was: a doctor, a teacher, or a cleaner.»[3] This leaves hope that artificial intelligence (AI) will not win of man, but there is no doubt that AI and the Internet change the environment and join with people in open competition. Being aimed at obtaining income or profit, or expansion of any business, artificial intelligence will quickly find that the easiest way to make money from money, blow bubbles, and also to earn on human vices and weaknesses.

Total, the cumulative result of the arrival of the digital economy is that the classical economic paradigm, that market equalizes the supply and demand by means of price, becomes to no adequate. The market mechanism works well everywhere, where supply is constrained by limited resources, the demand – limited income, and when preferences are stable. Until recently, these conditions have met almost everywhere, and the action of the free market applies to all major categories of economic agents (EA): low-income, middle-income and high-yield. In the digital economy the scale of financial bubbles such that:

– the low-income category expands, and by necessity, becomes a category of social dependents who live on benefits and subsidies do not have the available funds to choose on the open market, making use of direct distribution, for which the market is just a smokescreen;
– average category gradually but steadily narrowed and remains the only refuge for the classical market paradigm;
– high-yield category, while remaining quite narrow, concentrating the largest share of financial capital, is turning into a financial bubble and completely ceases to obey classical economic paradigm.

The concentration of capital in high-yield categories quite well characterized by the following figures: "Only 13% of the world's population own 95% of the capital of the Earth"[4], «1% of the richest people on the planet have assets in the amount of $110 trillion. Is 65 times larger than the combined volume of wealth that belongs to the poorest half of humanity.»[5] According to the famous Italian journalist and analyst Juliet Chiesa "This world is ruled by 9 people"[6].

---

[2] http://tvkultura.ru/video/show/brand_id/32318

[3] http://tvkultura.ru/video/show/brand_id/32318

[4] http://www.234555.ru/publ/11-1-0-70

[5] http://sunnapress.com/news/business/6447-85-samykh-bogatykh-liudei-v-mire-vladeiut-takim-zhe-obemom-aktivov-kak-i-polovina-naseleniia-zemli.html

[6] http://nvdaily.ru/info/38411.html

Economic science was well studied of the processes occurring in low-income and middle-income areas, that is where the classical paradigm is valid given the social support measures and direct distribution. However, the most interesting and absolutely unknown for science going on in high-yield category. It is home to consumers with the so level of solvency that they have access to any resources in any quantity. Their preferences are unstable. In other words, it is economy of unlimited supply and uncertain demand, in which there is no sense to speak of the equilibrium of demand and supply: the classical paradigm does not work here! Examples of the inadequacy of the classic paradigm are presented in numerous studies, including Nobel prize winners [8]. Mostly these are examples of imbalance due to psychological factors. These studies provide useful recommendations to developers of marketing programs, advise on how to generate and distribute reward funds among top managers of large companies, but they give nothing for understanding the processes inherent in the digital economy.

Scientific basis for solving the tasks are included the theory of Equilibrium stochastic processes (ESP) [1, 2], the methodology of mathematical modeling of ESP, namely Evolutionary-simulation methodology (ESM) [1, 2], measurement of financial bubbles [3], target of management by bubbles [3], an instrumental system "Decision" that allows you to programmatically implement ESM [4, 5].

In particular, in the theory of RSP introduced the concept of "Risk of overstatement" and "Risk of understatement" and proven (see Chap. 3 in [2]), that the balance of risks, that is, the equality:

$$Risk\ of\ overstatement = Risk\ of\ understatement$$

is a generalization of the equilibrium of supply and demand:

$$Demand = Offer$$

The balance of risks expresses the minimax strategy of EA that works in all situations, and when the EA no resource constraints, and when it has a changeable preferences and, when uncertain intentions. In all circumstances, the EA focuses on their subjective risks on the basis of how not to do something unnecessary (risk of overstatement) and not to miss anything necessary (risk of understating). So EA in all categories, low-income, middle-income and high-yield are focused on the risks. The balance of risks expresses a new paradigm, namely the paradigm of the digital economy. The classical paradigm of equilibrium of supply and demand is its special case.

The economy is formed by a combination of economic agents (EA). In the four-sector model (see Fig. 1) four categories of EA are distinguished: households (individuals and their families); firms (organizations aimed at the production and sale of goods and services); state (a set of institutions that provide for the regulation of the economy); foreign sector (all other states). We use the concept of EA in a broader sense, including here besides the mentioned 4 categories, as well as industries, spheres of activity, market sectors, regions and authorities.

EA is active, that is, it reacts to random external influences, seeking, at the same time, to achieve a certain goal. For example, a firm seeks to make a profit or income, or capitalization, or at least not to go bankrupt, to maintain an identity. The reaction of the

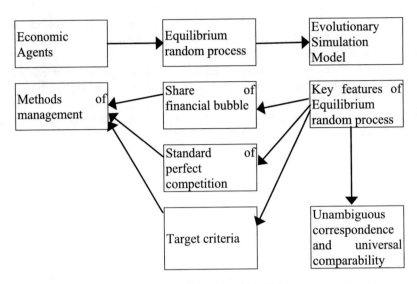

Fig. 1. The main components of the financial bubble management system.

EA to the external medium is called the control action. Accident and activity, combined, lead to the fact that the functioning of any EA is Equilibrium random process (ERP) - a process whose trajectory in the phase space is determined by a combination of random factors and control actions, the direction and force of which is determined by the size and direction of the deviation of the actual trajectory from smoothened [6]. Clearly, the meaning of ERP is seen in the example of the market of a particular product, where the phase coordinates are the price and the sales volume; the smoothed trajectory is the equilibrium price and sales volume; The actual trajectory is an instantaneous value of the price and volume of sales; Control actions are profit (for the supplier) and expenses (for the buyer). Evolutionary Simulation Model is also a model of the ERP and an EA model of any category.

The concept of "The share of the financial bubble" (SFB) was introduced in [5]. The SFB measures the share which is provided the secured portion in the nominal value of currency or securities. At the same time the SFB expresses the probability of the reliability of economic indicators, such as profit, market capitalization, prices, etc. SFB can be likened to the measure of the "strength of alcohol drinks", which is the ratio of "... the amount of dissolved anhydrous alcohol to the volume of the drink multiplied by 100%." (https://ru.wikipedia.org/wiki/Strength_of_the_drink). For example, if SFB of dollar is 97% [5], it means that the real price of the green paper is 3 cents. The remaining 97% - is bubble, financial nothing, emptiness. Along with this, the SFB expresses the probability of the reliability of prices or other economic indicator. For example, if capitalization of a company are equal N, then the SFB is the probability, that the figure N is true.

The method of calculation of the SFB based on the fact that the main characteristics of ESP, namely, the equilibrium value of the integral characteristics PL, the probability that the actual value will exceed the equilibrium (reliability) P^0 and the ratio of the risk of overstating and the risk of understating (O/U) in a neighborhood of PL mutually uniquely related, and the indicators $P^0, O/U, D$, where D- is the average proportion of risk-adjusted performance, is comparable. To calculate the SFB for specific EA (or currencies, or securities) must:

- First, find a reference EA (or currencies, or securities), which is similar to the studied EA and works in conditions close to perfect competition.
- Second, to calculate for the standard EA and for the tested EA (or currencies, or securities) at least one of the characteristics $PL, P^0, O/U$, meaning that for the calculation can often be used not only specially designed and implemented in the "Decision" ESM, but also a variety of straightforward ways. In particular, P^0 it is often possible to directly measure according to statistical reports or based on a special, restricted, confidential surveys. In many cases, you can use ESM set out in [2–5], which are already quite large library.
- Thirdly, knowing P^0 and 3/3, can be calculated $D = \frac{P^0}{O/U}$, and knowing D_S for the standard and D_I for the investigated EA (or currencies, or securities), SFB can be calculated by the formula: $SFB = \frac{D_S}{D_I} * 100\%$.

The purpose of management is the alignment of the SFB in different EA, but not the desire to ensure strict and continuous equality. It is only necessary to ensure the elimination of unwanted anomalies in the size and stability of bubbles. In sufficient detail management objectives are analyzed in [3]. According to proven theorems (see Chap. 3 in [2]) the bubbles increase the differences in values D and concentration of capital, where D more. Growth D contributes to economic growth, that is, to increase the capitalization of EA, or volume of production, or to improve stock prices, etc., and to the growth of SFB. Economic growth can be considered qualitative only if it occurs either at a constant SFB, or when its moderate increase, in any case, less than the growth of the economy. If the growth mainly is due to the SFB, it is not quality.

What, for example, make sense to discuss a great long-term growth in U.S. GDP and poke this growth our monetary authority if we do not know what proportion of this growth is the increase in a financial bubble? It may even be that it's not any growth, and net financial bubble and this growth does not need to be proud of, and it must actively fight. It is known that investments in infrastructure, education, new technologies offer slower GDP growth, which, moreover, are not fully reflected in the statistics. In contrast, all financial bubble is a direct contribution to the GDP of the country which resident is the owner. Recently introduced the game "Pokemon Go". "... Nintendo, which owns the brand Pokemon earned in a week more money than the Central Bank prints for the year"[7]. Nintendo, is typical for the digital economy. He does not need anything except the Internet: metal, energy, work. The question is, what Pokemon is better financial pyramid? Is it permissible through games so fast to concentrate such huge funds? For example, on the stock exchange, in the case of large jumps or panic suspend trading. Why not halt the spread of the game? Or not to impose special taxation? Or not to limit the ways of use of funds in the bubbles? Or not to

---

[7] http://07kbr.ru/2016/07/21/pokemon-go-v-rossii-na-soveshhanii-v-centrobanke-nashli-pokemonov/

restrict the movement of such funds? All these measures for control of bubbles, like Nintendo. An example from Nintendo is not a rare exception. Today bloggers formed millions of audience and make billions.

The task of monitoring the SFB should be assigned to the statistical service, the rating agencies and the Central Bank respectively the directions of their activities. They should be required to openly and regularly publish data on SFB. They have to develop and approve the calculation methods and use of SFB in the rating of cost of business, value assessments of currency quotes, the implementation of comparative studies. It is essential that methods of calculation of the SFB, the sources of information for calculation and the software was also available for free (at public expense) with online access, which should be enshrined in law. Everyone should be able opportunity to quickly, in a dialogue mode calculate SFB for yourself, for the competitor, for the transaction, etc. Otherwise, the calculation of the SFB, the SFB assignment for the business or securities, immediately become a method of unfair competition, pressure, blackmail and, ultimately, will be another way of concealment, preservation and inflating bubbles.

3 Conclusion

Summarizing all above, we come to following conclusion: it is necessary to propose the paradigm of digital economy, which will cover all categories of economic agents and are therefore a generalization of the classical market paradigm, and will enable to develop technology for the management of financial bubbles, including:

- methods of measurement of bubbles;
- the target criteria to the control of bubbles;
- ways of influencing bubbles.

The purpose of managing a financial bubble is defined by the fact that:

(1) management of bubbles systematically solves a set of interrelated and complementary objectives, including the achievement of economic justice (a concept introduced in [5]), increasing the reliability and accuracy of economic measurement, reducing disparities, improving efficiency;
(2) provides the possibility of effective state intervention in the economy no significant constraints to the mechanisms of market self-regulation, but on the contrary, relying on these mechanisms;
(3) in the digital economy the importance of the management of bubbles increases rapidly and gets, apparently, of paramount importance.

References

1. Avdiyskiy, V.I., Bezdenezhnykh, V.M., Liechtenstein, V.E., Ross, G.V.: Economic equity and security of economic agents. Finance and Statistics (2016)
2. Fry, J., Brint, A.: Bubbles, blind-spots and brexit. Risks **5**(3), 37 (2017). http://dx.doi.org. aucklandlibraries.idm.oclc.org/10.3390/risks5030037
3. Guerrero, A., Hawser, A., Platt, G.: Trouble with bubbles. Global Financ. **28**(9), 44-54, 56,58,60-62,64,66-67 (2014). http://0-search.proquest.com.www.elgar.govt.nz/docview/1617 154729?accountid=40858
4. Lichtenstein, V.E., Ross, G.V.: Equilibrium stochastic processes: theory, practice, infobusiness. Finance and Statistics (2015)
5. Lichtenstein, V.E., Ross, G.: New issues in the economy. Finance and Statistics (2013)
6. Lichtenstein, V.E., Ross, G.V.: Information technologies in business: application of "Decision" in micro- and macroeconomics. Finance and Statistics (2008)
7. Lichtenstein, V.E., Ross, G.V.: Information technologies in business: application of "Decision" in solving applied economic problems. Finance and Statistics (2009)
8. Lichtenstein, V.E., Ross, G.V.: Nobel prize and the ideology of economic justice. Economic and Human Sciences, No. 5(292) (2016)

Informational Technologies for the Efficiency of Public Debt Management in Russia

Svetlana Tsvirko$^{(\boxtimes)}$ ⓘD

Department of World Economy and World Finance,
Financial University under the Government of the Russian Federation,
49 Leningradskiy prospekt, Moscow, Russian Federation
s_ts@mail.ru

Abstract. The article focuses on the significance of information technologies for increasing government efficiency in public debt management in Russia. This paper reveals the main problems of Russia's public debt management. We have suggested directions of improvement the system of public debt management. International experience in building comprehensive computer-based debt management systems has been summarized. Such aspects of information systems for public debt management as functions and tasks, properties and main elements were discussed.

Keywords: Public debt management · Efficiency · Informational technologies
Macroeconomic indicators · e-government

1 Introduction

Public debt management is one of the most important directions of economic policy and a constant search of its efficiency is one of the purposes of the government.

Managerial decisions should be based on promptly collected and processed relevant information. For this reason the use of information technology is considered to be one of the possible mechanisms to improve the efficiency of public administration at the present stage. Information technologies allow to reduce the time necessary for the implementation of the processes within the government body. This is also accompanied by a direct increase in the economic efficiency of the state body (for example, through strengthened labor productivity, reduction of costs, etc.).

There is significant amount of publications devoted to the problem of public debt management – both in western economic literature and in Russia. Among them we can name publications by foreign authors, such as Reinhart et al. [1], Cecchetti et al. [2], Kos [3], and many others and research carried out by Krasavina [4], Kheyfets [5], Zvonova [6] in Russia. At the same time there are many publications connected with informational technologies in common. But there is a lack of publications devoted to the features of modeling and application of information technologies in public debt management in Russia.

In this context the aim of this study is to analyze application of informational technologies in the sphere of Russia's public debt. For achieving the results of the research, it is necessary to clarify the problems of the system of the public debt management in Russia,

© Springer International Publishing AG, part of Springer Nature 2018
T. Antipova and Á. Rocha (Eds.): MosITS 2017, AISC 724, pp. 104–113, 2018.
https://doi.org/10.1007/978-3-319-74980-8_10

name the directions of improving the system of public debt management and the peculiarities of applications of informational technologies in this sphere.

2   System of the Public Debt Management in the Russian Federation

The system of the public debt management is understood as the interrelation of budgetary, financial, accounting, organizational and other procedures aimed at effective regulation of public debt and reducing the impact of debt burden on the economy of the country. The system of public debt management includes strategic and current (operational) management. Strategic management is associated with setting the main benchmarks for the public debt and creating the institutional and legal mechanisms necessary to achieve the goals set in this field. The system of debt management comprises subsystems such as the following: infrastructure (institutional subsystem); legal support (legal subsystem); economic operational regulation (subsystem of debt parameters); technology (trading subsystem); debt settlement system.

Traditionally, the main objectives of public debt management are as follows:

- maintaining debt at an economically safe level;
- ensuring the timely execution of debt obligations in full;
- minimizing the cost of debt.

Generally, the above mentioned objectives were achievable for Russia. As of the 1st of 2017 the indicator of domestic and external public debt was 13% of GDP, that is quite low [7]. The situation in the Russian Federation with the public debt some time ago seemed quite prosperous. However, under the conditions of limited budget possibilities, the task of forming an effective debt strategy of the state is actualized. There are studies that clearly demonstrate that there are problems in the sphere of public finances of the Russian Federation. For example, in the report "Russia's Fiscal Gap" the experts noted that there is a significant budget gap in Russia: the accumulated difference between the current value of all future expenditures and budget revenues is about 15 annual GDP (in year 2013 prices). The liquidation of the budget gap requires an annual tightening of the budgetary policy: an immediate and constant increase in taxes by 29%, or a regular reduction in spending by 22.4%, or a combination of these two measures [8, p. 2]. If we postpone the tightening of budgetary policy, then the excess of budget obligations over revenues will only increase, and the problem will be aggravated. Thus, the study clearly shows that Russia is facing a serious long-term budget problem.

The growth of the public debt of the Russian Federation was accompanied by the surplus of the budget in year 2011 and relatively small deficit in year 2012. As the result of the analysis it was found out that above mentioned situation occurred for the following reasons:

- as a result of the mechanism for the formation of the Reserve Fund, when part of the oil and gas revenues received during the execution of the federal budget is not spent in the current fiscal year and is sent to the Reserve Fund;

– due to an inaccurate forecast of oil and gas revenues due to the underestimation of projected price of oil.

The deficit of the federal budget is financed by borrowed funds, that, in turn, leads to an increase in the public debt of the Russian Federation and the costs of its servicing.

At the same time, the weighted average cost of borrowing is 3–7.5 times higher than the average yield from the investment of the sovereign funds such as the Reserve Fund and the National Welfare Fund of the Russian Federation.

It can be argued that, in spite of the formally low indicators of the Russian government debt and some past successes, in particular, a significant reduction in the state external debt in the 2000s years, on the whole, Russia's public debt management system is inefficient; actions to manage public debt and sovereign funds are not consistent with each other.

Public debt would have been much higher in case of the inclusion of the quasi-sovereign debt. Formally, the Government of Russia is not responsible for the debts of state companies and banks, but investors and rating agencies expect that such companies will be supported if necessary.

Already for a long period of time the expert community has been noting that it is urgent to monitor the situation with debts in terms of their high concentration within the small number of borrowers, as well as a sharp increase in quasi-sovereign debt. There is an urgent need to develop measures to prevent the growth of such debt, but to date no real mechanisms have been developed to limit and control quasi-sovereign debt.

An important aspect of public debt and its management is the problem of so-called contingent liabilities. The contingent liabilities of the budget represent a kind of financial category that is absent in the Budget Code (they are not implicit, or conditional budgetary obligations). These are obligations that arise in the state under certain circumstances in the future. Based on the experience of foreign countries, experts name examples of conditional obligations, such as: state guarantees and public insurance systems; pension liabilities of future periods; obligations of regional and local authorities; obligations within the framework of public-private partnership; liabilities of corporations with state participation; obligations to maintain stability of the financial system. There is a steady growth of contingent liabilities in the world, including the leading developed countries. The problem of contingent liabilities is also relevant for Russia. Unrecorded risks associated with Russia's debt may lead to the need for costly anti-crisis measures that will require state budget funds.

Once again, we emphasize that there is no complete comprehensive accounting of all contingent liabilities accumulated by governments, which leads to the lack of the authorities' ability to manage contingent liabilities and emerging risks.

The main problem in the sphere of public and private debt is the lack of opportunities for sustainable debt refinancing through domestic and external borrowing in the required quantities and at favorable terms and conditions.

Thus, Russia's system of public debt management has a number of unresolved issues, both methodical and practical.

3 Directions of Improving the System of Public Debt Management and the Application of Informational Technologies

There is a need to expand the objectives of public debt management. They should be supplemented with the following:

- ensuring stable servicing of both external and internal state obligations in any crisis situation, monitoring the situation with private external debt;
- flexible response to the changing conditions of external and internal financial markets and the use of the most favorable sources and forms of borrowing;
- strengthening the attractiveness of Russian public and private borrowers, preventing sharp fluctuations in the price of liabilities on the global financial market;
- forecasting and avoiding risks related to the debt structure;
- coordination of public policy and external borrowing policies of corporations in order to avoid unnecessary competition in financial markets and risks associated with possible non-fulfillment of corporate obligations;
- creation of an integrated, effective system of public debt management;
- improving the accounting and monitoring of public debt, the introduction of advanced technologies for debt management, allowing monitoring of the debt burden and of the obligations' fulfillment in real time.

In order to ensure macroeconomic stability, countries use different approaches to control the economic situation and debt policy mechanisms. This results in the application of modern methods to address the problem of sustainability of public debt.

Information technologies related to public debt allow us to carry out a comprehensive analysis of the problem under investigation. With the help of information technologies, it is possible to adequately describe and assess causation of public debt and find its safe level, which does not threaten the stable development and welfare of the state.

According to the Guide for compliers and users «Public sector debt statistics» , a debt office should store information in an efficient and comprehensive computer-based debt management system (CBDMS) that can undertake a number of tasks and so support both operational and policy functions. Typically, a CBDMS should be able to provide:

- debt recording;
- debt reporting;
- debt analysis;
- linkages with other packages and systems of the public sector unit [9, p. 124].

Several information systems for public debt management are known on the international market. They have some differences, connected with the tasks to be solved and the functions performed. It is necessary to name the following informational systems:

- UNCTAD's debt management and financial analysis system (DFMAS) and
- The Commonwealth Secretariat's software suite for debt management (DFMAS).

Since the early 1980's years UNCTAD's programme "Debt Management and Financial Analysis System" (DMFAS) has been successfully used in public management. The programme is available to all countries on their request. So far it has been used directly at the country level in 69 countries (mostly low and lower-middle income) and is currently in use in 86 institutions in 57 countries. The DMFAS programme is a computer system designed to solve the tasks of public debt management in ministries of finance and/or central banks. It provides users with the ability to monitor short-term, medium-term and long-term debt, both internal and external. The system can also take into account private debt and grants, as well as lending operations.

The current version of DMFAS 6 was launched in November 2009. DMFAS 6 has a web-based interface (portal) that provides centralized access to all the modules, information, applications, data and links that are normally used by its users in DMFAS. It also contains an on-line help system.

DMFAS can be used as a stand-alone system or in a network (intra-departmental or external). User profiles and access rights are defined in the system security module. Simple and independent access to all modules with customization for specific user requirements is provided. Modules are placed in such an order that corresponds to a typical cycle of a debt agreement (administrative tasks, mobilization and debt servicing) with addition of such functions as agreement on terms (in the case of debt securities), reports and analysis. The modules cover all the needs of the debt management department, including the tasks of the operations department (issuing debt securities), middle office (analysis) and back-office (registration and operations' management).

DMFAS 6 works through any standard Internet browser. It uses Oracle's relational database management system and it was developed using the Java and Oracle tools [10].

The Secretariat of the Commonwealth of Nations offers advice on debt policy and technical assistance. The Debt Management Section of the Special Advisory Services Department carries out activities aimed at improving the accounting, monitoring, management, analysis of all categories of debt. Thus, the CS-DRMS software product was developed – it is an integrated tool that allows registration, monitoring, analysis and reporting on public debt.

The CS-DRMS system was established in 1985 for Sri Lanka, a member of the Commonwealth and experiencing problems with managing external debt. Initially, the CS-DRMS gave an opportunity of the accounting and management of external debt. Taking into account the need to manage internal public debt as well, appropriate additions were made to the system.

At present in addition to fulfilling the functions of an integrated repository of information on all types of external and internal debt (including guarantees) for each instrument separately and in aggregate form, this system allows users also to keep records and manage the debt of private sector and grants. The system can be used for the whole cycle of debt operations from the moment of agreeing on the conditions for obtaining a loan or issuing a security and until the redemption of the debt.

The system has a user-friendly graphical interface that allows the debt manager to obtain quickly information about government borrowings and related transactions using various aggregation facilities. A holistic approach involves a wide range of functionalities and a reliable, secure and open architecture with interfaces to multiple external systems.

The CS-DRMS system includes fully customizable multi-level security tools that allow configuration of workgroups and distribution of roles. This makes it possible to adapt to the scheme accepted in the country with the distribution of goals and tasks of the front office, middle office and back office. Users can be assigned to appropriate roles with restricted access to certain screens and reports in accordance with the security system configuration.

The CS-DRMS has a client-server architecture, runs on the Windows platform. Software products work on both Oracle and Microsoft SQL Server systems as server databases. The software is available in English and French and is built in a language-independent manner to facilitate its translation into other languages.

The CS-DRMS is provided to the Commonwealth Member States for free, and can be purchased by other countries. Currently, it is used in more than 60 countries (including 15 non-Commonwealth members) on more than 100 platforms of ministries of finance, central banks, treasuries, regional governments. It manages a global portfolio of more than $2.5 trillion of public debt [11].

As we can see, a large number of countries and institutions use ready-made information systems related to public debt. But also a large number of countries create their own solutions for the relevant areas (mostly developed countries).

The advantages and disadvantages of the information system for public debt management that is acquired or created independently are analogous to this situation with any information system in business.

The advantages of buying of an information system is fast delivery, significant amount of functions with a variety of types of reports. Purchase is cheaper than creation of your own information system. The drawbacks in this case are the receipt of unnecessary functions of the system and the lack of necessary ones; complexity (and sometimes impossibility) of self-development of modules, that are needed.

The independently developed information system is much better suited to goals and objectives. The reports are more individual and the final results will be more successful. Deficiencies in the development of the information system are longer periods and higher costs of creation. At the same time, we should not forget that, ultimately, the creation of an information system for public debt management has aspects of economic and information security of the country.

The considered information technology of public debt research are characterized by the following properties:

- reliability and completeness of information included in the structured database of information technology, on the basis of which further analysis of the public debt is carried out;
- conformity of the information technology data to the time of the research;
- access to information of this information system;
- sufficient quantity and completeness of components and structure of information technology;
- the feasibility of constructing of information system.

The information technology of public debt management allows to analyze the corresponding type of debt and macroeconomic indicators that influence the dynamics of the latter and allows to assess the changes of the public debt in the time period in accordance with external and internal economic factors.

Opportunities provided by the information technology system of public debt should be as follows:

- systematization of statistical data;
- finding of "key" factors of influence on the dynamics of the public debt;
- allocation of necessary economic indicators;
- conducting analysis, estimation and forecasting of the public debt;
- carrying out of analysis and estimation of received forecast data.

Tasks solved by information technologies in the sphere of public debt can be formulated as follows:

- identification of elements of the structure of the public debt and their economic essence;
- grouping of sources of borrowed funds and sources of repayment;
- identification of the consequences of indebtedness on economic processes;
- implementing a comprehensive statistical analysis of the public debt;
- study of changes in the main indicators characterizing the public debt;
- calculation of the forecast value of the public debt;
- analysis and evaluation of the received data, formulation of recommendations.

The development of the he information technology for public debt management should include several stages. The main stages are structuring of the database of public debt and macroeconomic factors, influencing debt sustainability; econometric modeling and building forecast model; analysis and evaluation of the public debt.

Each step can be described in more detail.

On the stage I it is necessary to structure the database, make the grouping of statistical information, formulate those macroeconomic indicators, that have impact on economic development and debt sustainability.

A study conducted by us and presented in some previous publications [12] showed that the following economic indicators are needed to formulate a mathematical model for analyzing and assessing of the public debt:

y_t - share of domestic public debt in GDP;

y_t^0 - share of external public debt in GDP;

x_t - rate of output growth;

π_t - inflation rate;

r_t - weighted average interest rate on domestic public debt;

r_t^0 - weighted average interest rate on external public debt;

δ_t - growth rate of the real exchange rate of a foreign currency in relation to the national currency;

β_t - share of external debt convertible into the national currency (which corresponds to some types of conversion, for example, debt/shares, debt/goods, debt/exports);

γ_t - share of the external public debt to be written off;

φ_t- share of the public debt, subject to reduction through the repurchase of debt from reserves - funds of sovereign funds;

$\lambda_t = g_t - \theta_t$ - relative primary deficit (the difference between budget expenditures and budget revenues);

λ_{ts} - share of the aggregate budget deficit in GDP;

α_t - share of the total budget deficit financed by the credit and money issue of the Central Bank.

The difference between this model and those suggested by other authors is that the parameters reflecting the use of debt settlement methods and related to the use of state reserves (sovereign funds) are taken into account. It gives an opportunity to make forecasts and analyze the dynamics of public debt with a wide range of possible conditions and scenarios.

On the stage II the patterns and trends of economic development are established. This is the stage of the development of econometric models. The selected econometric models should accurately reflect the trend changes. At this stage macroeconomic indicators should be studied. Correlation analysis of macroeconomic indicators should be performed. It is necessary to identify the interrelations between variables and exclude those that are linearly dependent (eliminate multicollinearity).

Stage III of the process includes assessment and analysis of the public debt, taking into account changes of variables. The next step is the forecast of the public debt and making recommendations for the actions in the corresponding sphere.

The economic and mathematical models that form the basis of information systems must be verified. As for informational system for public debt management as a whole it should be regularly audited [13].

4   Recent Developments and Recommendations for the Improvement of Public Debt Management in Russia

By the Order of the Government of the Russian Federation of July 20, 2011, the Concept for the creation and development of the state integrated public finance management information system "Electronic Budget" was approved. Then, by the Decision of June 30, 2015, the Government of the Russian Federation approved the Standing Order on the "Electronic Budget" system [14]. The "Electronic Budget" is intended to ensure transparency, openness, accountability of the activities of state bodies and management bodies of state non-budgetary funds, local governments, state and municipal institutions and to improve the quality of their financial management. In 2015–2017 years it was planned to develop subsystems for managing revenues, expenditures, cash, debt, financial assets, accounting and reporting, financial control and information and analytical support. For the years 2018–2020 it is planned to develop a subsystem for managing non-financial assets, human resources, as well as the development of a single portal of the budget system of the Russian Federation and subsystems introduced earlier.

The work on the application of accounting and reporting standards in the public management sector of the Russian Federation was initiated by the Ministry of Finance of the Russian Federation and the Federal Treasury, but is in the process of improvement. It is required to enhance accounting in the public management sector within the framework of the project "Modernization of the Treasury System of the Russian Federation". So far there is a lack of coordination of informatization processes conducted at various levels. As a result, state databases are partially overlapping each other, they are created based on various algorithms, that exclude the possibility of exchanging information between them in real time. Different systems should be integrated in a conceptual plan, their architecture should provide the possibility of communication between different systems for monitoring budget implementation.

Under current situation, the following recommendations could be made:

1. To increase the quality of analysis and forecasting, it is necessary to improve the model tools and decision-making methods. Economic-mathematical modeling should be activated. When appropriate it is advisable to use the experience and advanced developments of international organizations, such as the World Bank, the International Monetary Fund, UNCTAD, the Bank for International Settlements and others, as well as central banks, debt agencies, financial agencies, sovereign funds, etc.
2. It is necessary to consider a set of admissible scenarios within predictive constructions. The reliability of the forecasts should be increased. The coordination of the general outlines of the forecast scenarios used by various ministries and departments is required.
3. It is advisable to strive for information transparency. This helps to improve understanding of the decisions in the corresponding sphere. It is necessary to take into account that, for example, scenario forecasts perform an important function of forming the expectations of society and economic entities for the corresponding period of time.
4. The condition for effective implementation of the budget policy is the broad involvement of experts and the society as a whole in the procedures for discussion and adoption of budgetary decisions, as well as public control.
5. It is necessary to strive for the transition from inertial forecasting on the basis of trends towards the design of the future.

The development of public debt management of the Russian Federation should be considered within the framework of the public management system as a whole. It is advisable to do this in the logic of the concepts of "new state management", "e-state", "e-government", "network management", "anti-crisis public management", which are modern concepts aimed at developing an effective public management system [15].

5   Conclusion

In this paper we have summarized the key problems of Russia's public debt management. We would like to stress, that accountability, transparency and interdepartmental coordination play significant role in public debt management.

The results show that modelling and usage of informational technologies contribute to the improvement of debt management. It is mathematical modeling and information technologies that provide an opportunity to analyze financial flows, make predictions and choose rational solutions. The models expand the apparatus of the quantitative forecast of the debt strategy of Russia under different conditions and scenarios. The practical importance of the research lies in the fact that its provisions and conclusions can be used to solve tasks within the debt strategy. The results of the study are oriented towards practical application by ministries and departments dealing with public finance issues (Ministry of Finance, Central Bank).

References

1. Reinhart, C., Reinhart, V., Rogoff, K.: Dealing with debt. J. Int. Econ. **96**(Supplement 1), S43–S55 (2015)
2. Cecchetti, S.G., Mohanty, M., Zampolli, F.: The real effects of debt. Working papers 352 (2011). www.bis.org/publ/work352.htm
3. Koc, F.: Sovereign Asset and Liability Management Framework for DMOs: What Do Country Experience Suggest? (2014). http://blog-pfm.imf.org/files/fatos-salm_final-edited.pdf
4. Gairadzy, S.I., Krasavina, L.N. (eds.): Debt Policy: World Experience and Russian Practice, 159 p. Financial University under the Government of the Russian Federation, Moscow (2011)
5. Kheyfets, B.A.: Risks of debt policy of Russia on the background of the global debt crisis. Issues Econ. **3**, 80–97 (2012)
6. Zvonova, E.A.: Current problems of public debt and sovereign default to the EU countries. Finance: Theory and Practice (previous name «Vestnik Finansovogo Universiteta» until 2017), vol. 20. 4 (94), pp. 105–117 (2016)
7. Ministry of Finance of the Russian Federation. https://www.minfin.ru/ru/perfomance/public_debt/
8. Goryunov, E., Kazakova, M., Kotlikoff, L.J., Mamedov, A., Nesterova, K., Nazarov, V., Grishina, E., Trunin, P., Shpenev, A.: Russia's Fiscal Gap. NBER Working Paper No. 19608 (2013)
9. Public sector debt statistics: guide for compliers and users. International Monetary Fund, Washington, DC (2011)
10. Debt Management and Financial Analysis System (DMFAS) Programme. http://unctad.org/divs/gds/dmfas/Pages/default.aspx
11. Public Debt Management Programme. http://thecommonwealth.org/public-debt-management-programme
12. Tsvirko, S.E.: Model support for the debt strategy of Russia. Econ. Entrepreneurship **6**(71), 787–792 (2016)
13. Guidance on Auditing Public Debt Management Information Systems (2016). http://www.wgpd.org.mx/Anexos/210217/GuidanceAPDIS_e.pdf
14. The tasks of the public finance management information system "Electronic Budget" are defined. http://www.garant.ru/news/635744/
15. Okhotsky, E.V.: State management: on the way to a modern model of public management. Bull. Moscow State Inst. Int. Relat. (MGIMO). **3**, 115–127 (2014)

Streamline Management of Arctic Shelf Industry

Tatiana Antipova$^{(\boxtimes)}$

Federal State Budgetary Educational Institution of Higher Education
"Perm State Agro-Technological University Named
After Academician D.N. Pryanishnikov", Perm, Russia
antipovatatianav@gmail.com

Abstract. The current research covers one of the most urgent management issues of today's world in Arctic Shelf industry on the Russian part. The research design is characterized as quantitative: data collection includes official annual and sustainable reports of biggest energy companies, articles, books and the other information resources. We analyzed how to modernize and develop the management and safety of the Arctic zone. Streamline management can led to high productivity and efficiency of the Arctic Shelf industry. Also we design lucrative scheme to increase benefits and decrease the cost of development the Arctic shelf industry. Arctic shelf development is a prerequisite for economic power of the country and here, an important factor in streamline management of production processes and reducing related production risks is competent. Russian Arctic zone generally is a huge resource reserve of the country and among the few world regions with practically untapped hydrocarbon, mineral, and others reserves for developing streamline management of Arctic Shelf industry.

Keywords: Streamline management · Reduction of the cost
Production output · Offset expenses · Streamline management
Arctic shelf industry

1 Introduction

For the countries that border the Arctic Ocean—Russia, the United States, Canada, Norway, and Denmark (through its territory of Greenland)—an accessible ocean means new opportunities. All five Arctic Ocean neighbors want to get slice of the Arctic. These countries quietly send in soldiers, spies, and scientists to collect information on one of the planet's most hostile pieces of real estate. The actual decision on which country will be held the Arctic balance established under Article 76 of the United Nations Convention of the Law of the Sea (UNCLOS). This treaty is a sort of international constitution establishing the rights and responsibilities for the use of the world's oceans including thousands of miles from the frigid North. Today, nearly 170 countries have ratified or acceded to the treaty, but the United States has yet to do so. In fact, out of the five Arctic Ocean nations, the United States is the only outlier [2]. The situation is complicated that even China claims the use of the Arctic. China is joining the Arctic due to its proximity and as a near state may be impacted by Arctic climate

© Springer International Publishing AG, part of Springer Nature 2018
T. Antipova and Á. Rocha (Eds.): MosITS 2017, AISC 724, pp. 114–121, 2018.
https://doi.org/10.1007/978-3-319-74980-8_11

change, justifying the need for research, due to economic opportunities with the opening of Arctic passages and, according to governmental sources, better Arctic governance (China Daily, 2013).

The Arctic is a complex place undergoing great change early in the twenty-first century, and there is much uncertainty and speculation about what the future might hold for this once remote and inhospitable region of the globe. One methodology that can enhance strategic planning in regard to the Arctic is the use of scenarios and the creation of plausible stories about the future [3].

This scenario relies on such key studies as the work of the Intergovernmental Panel on Climate Change (for global climate projections); the Arctic Council's Arctic Climate Impact Assessment (2004); scenarios developed in the Council's Arctic Marine Shipping Assessment (2009); and the EU's Arctic, Climate Change, Economy and Society (ACCESS) project (2014) [9]. For these scenarios realization we need to understand where are the Arctic zone borders. This question is discussed on present day. On Fig. 1 show approximately borders and we can see that some of its still uncertainty and controversial.

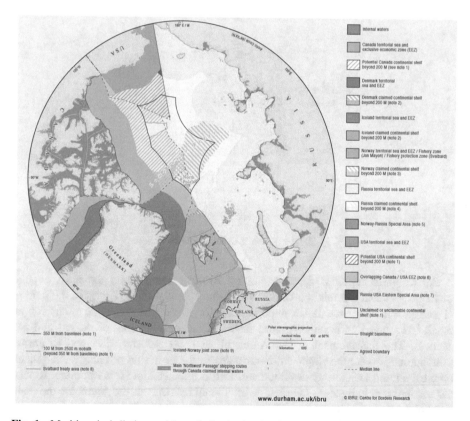

Fig. 1. Maritime jurisdiction and boundaries in the Arctic region. Source: http://www.durham. ac.uk/ibru

One of the most important reasons for this is that "petroleum has always been an important ingredient for geopolitical conflict and socio-environmental damage. The current international relations system as well as the international law does not address effectively the oil and gas political issues or the liability for damages from drilling activities" [1].

2   Research Design

The research design is characterized as quantitative: data collection includes official annual and sustainable reports of biggest energy companies, articles, books and the other information resources. According to literature review result, Arctic Shelf Industry can be dividing on four directions as shown on Fig. 2.

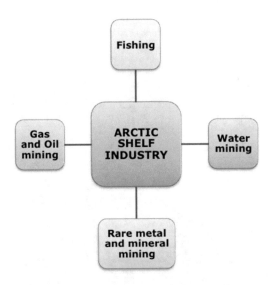

Fig. 2. Arctic Shelf Industry classification.

All of these activities decrease natural Arctic resources and the most popular activity is Gas and Oil mining. Undoubtedly, the potential of the Arctic in the field of oil and gas industry opens up opportunities for maximum economic profit for all stakeholders and this is an important factor in the development of the oil and gas business. The management of the Arctic oil and gas industry, designed to ensure economic and social development, contributes to the provision of energy equipment, job creation and technology development. But in the meantime it is important to take into account that indigenous peoples, as traditional residents of the Arctic are the main stakeholder and an integral part of any development, evaluation and strategy for the region. The work of oil and gas companies is as socially difficult as it is technically complex.

When relating the evaluated resources (fields) to the nearest country with equipartition of "conjoint" hydrocarbon-bearing fields, according to estimates, more then half of the aggregated resources of Arctic are in Russian segment covers the major reserves of natural gas, whilst major reserves of oil are in USA (Alaska) Arctic segment. So it is logical to consider Russian Arctic Shelf Gas and Oil industry. The study examines Russian law and official annual reports of the Russian Public Joint Stock companies working in the Arctic region.

3 Research Result

Russia has long had ambitions to produce large volumes of oil and gas from the Arctic. Currently, in the Russian Arctic produce 380 million tons of oil and 500 billion cubic meters of gas, that share in global consumptions 9.74% and 16.7% respectively.

It should be noted that only two companies are admitted on the Arctic Shelf, with shelf works experience for at least five years - Gazprom and Rosneft, having respective licenses. Gazprom and Rosneft have of most licenses, mainly for works in the Okhotsk, Kara and Barents seas. However, according to the program of the Russia Arctic shelf development, the right to conduct exploration and production of oil and gas in coastal waters may be also granted to other companies, including some smaller, private companies or subsidiaries of public enterprises.

The participation of foreign investors in mineral exploration, including the Arctic North, is limited by the Federal law No. 58 dated 29.04.2008, which exclude the participation of foreign companies, allowing their joint activity only with Gazprom and Rosneft [5].

Let consider and compare legal and business status both firms.

Gazprom. In accordance with Gazprom Charter, p.1.1., the joint-stock company Gazprom (hereinafter Gazprom) is a public joint-stock company. Gazprom founder is the Government of the Russian Federation (p. 1.2., Gazprom Charter). But this founder does not have any property in Gazprom because Gazprom is the owner of the property (p. 4.2, Charter of Gazprom), not Russian Federation.

Rosneft. In accordance with Rosneft Charter, Public (Open) Joint Stock Company "Oil Company "Rosneft" (hereinafter - Rosneft) was established in accordance with the Presidential Decree No 327 dated 01.04.1995 "On urgent measures to improve the activities of oil companies" and by order of the Russian Federation Government No 971 dated 29.09.1995 "On the transformation of the state enterprise "Rosneft" to the Open Joint Stock company "Oil company "Rosneft". Rosneft is the successor of the reorganized "Rosneft" state-owned enterprise in accordance with the transfer act.

As a comparison result we can reach the conclusion that both Gazprom and Rosneft are ordinary commercial Join Stock companies, not state-controlled institutions. In open access there is no truly information about share of Government stakes. This information can be obtained only by law enforcement agencies. According to some online reports it can be assumed that the state owns about 25% stake in Gazprom and Rosneft. There are no any mentions of this in the constituent documents of Gazprom and Rosneft. Anyway 25% is not controlled stake. In addition, Russian law does not define what is a state-controlled institution. Moreover both companies' main goal is profit. Unclearly, why someone is believed that Gazprom and Rosneft are state-controlled (public) institutions.

Respective comparative Annual progress reports of Gazprom and Rosneft were used official sites of the companies (http://www.gazprom.ru, http://www.rosneft.ru), reports of ecological companies (e.g. reports of Greenpeace). It is necessary to note that name of key financial indicator in English are different in annual Gazprom and Rosneft reports. These indicators are compared in equal meaning of Russian reporting due to author experience in commercial accounting. Comparative analysis of main key financial indicator set in Table 1.

Table 1. Key financial result indicators comparison, bln Rub.

Indicators	Gazprom			Rosneft		
	2014	2015	2016	2014	2015	2016
Net sales revenue	5590	6073	6111	5503	5150	4988
Tax	920	925	1188	3100	2300	2000
Net profit	1310	1228	726	350	356	201
Free cash flow	654	390	202	596	744	302

Table 1 data shows that dynamic of Gazprom and Rosneft amount of indicators is different. For instance, last three years Gazprom Net sales revenue is increased but Rosneft Net sales revenue is decreased. But it means nothing because Net sales revenue includes affiliates and associates firms belonging to main office as Gazprom as Rosneft. In some case indicators of several affiliates do not including in consolidate report caused avoiding taxation. It is explain why Tax amount so unpredictable different year by year. The same meaning has Net profit behavior. As a result of streamline management author draw the lucrative scheme to increase truly Net sales revenue and decrease the cost of the Arctic shelf industry.

To make this business more profitable usually using some lucrative schemes to streamline taxation in Russia. For this there are using minimum two legal models. These models were used by author being as chief of accountant at several commercial enterprises. The essence of these models is:

1. To create two or more subsidiary firms for dividing revenue and expenses flows. These must not be explicitly affiliates. Most of the revenue reflects in accounting and reporting in one firm (preferable with state support, to show budget money spending), but most expenses - in another firm.
2. To increase purchase prices for materials, raw materials, tools, etc. by passing overheads through several intermediate firms. Thereby increasing costs in the parent company and the decrease in income of subsidiary companies.

Still the cost of development and hydrocarbon extraction on the Arctic shelf is considerable. Reduction of the cost of oil and gas reserves developing in Arctic is possible, first of all, by production output of the most large deposits, which could partially offset expenses for a smaller oilfield construction. Appraisal of the oil-gas

Projects include most economical efficiency (internal rate of return, net present value, etc.) At present, most of the investigations prove that extensive development of Arctic shelf oil-gas reserves require huge investments. At that, return of investment, could be protracted for years.

Additionally, considering the management of the Arctic industry, we must take into account the importance of maintaining a balance between the extraction and conservation of natural resources, since we must think about our children and grandchildren. What will they get with the unlimited destruction of the wealth of the Arctic bowels? To consolidate this process accurately, the oil and gas industry will need to enhance a widely shared understanding of the present Arctic context under scientific, political and legal perspectives. The industry will need to clearly identify the risks and impacts of Arctic offshore development as well as demonstrate that those risks and impacts are being properly managed and mitigated. This is a challenge in itself.

Another fundamental aspect is to dominate the range of technologies to minimize the impact of oil and gas operations on the delicate Arctic environment. Technological innovation must be accompanied by transparency and ethics making clear to Arctic and non-arctic nations and Indigenous Communities, accurate risk assessment and safety procedures for Arctic operations. It will be essential to demonstrate that the industry can operate in the Arctic Ocean safely and sustainably. This will require a level of dialogue, cooperation and transparency that the industry has not achieved so far [1].

We support the idea of creating major strategic transport hubs in the North of Russia and deploying public logistics centers that lie at the intersection of the most important transport routes used by the key means of transport (particularly maritime ports located in the estuaries of the major Russian rivers), including the comprehensive deployment of air, rail, road, and pipeline transport.

Finally, it is worth noting that it is important to assess the multiplying effects of investment in transport and military infrastructure in the Arctic region, which will propel growth in heavy machine, equipment, and shipbuilding, and indirectly in metallurgy and other sectors of the economy [6].

It is reasonable to apply the real options method to implementing of streamline management when the following conditions are complied with:

- The result of the activity undergoes a high level of indefiniteness
- The company management can take flexible management decisions when new data on the reports occurs, and
- To a great degree the financial result of the project depends on the decisions taken by the management.

The proposed scheme of the streamline management involves six simple steps that can go a long way toward ensuring success and sustained demand streamline management process (Fig. 3).

The significance of the offshore development activities for Arctic shelf industry justifies the need for a multilateral assessment of their quantitative and qualitative indicators, including the possibility of achieving economic, social, political and innovative effects.

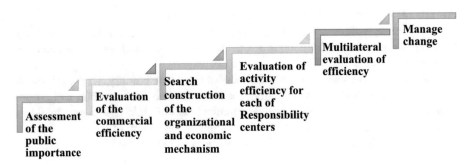

Fig. 3. The proposed scheme of the streamline management.

Besides, the Arctic regions, which are the subjects of Russia, are of substantial significance for the Country economy due to the fact, that its direct contribution to GDP is about 15% [7], and standard of living and natural resources potential by "quality of life" index are fairly high according to the results of Economy Institute Uro RAN investigations.

Thus, importance of the North and Arctic Regions for the present-day Russia is due to the fact that, on one side – it is extensive vital part of the Country's territory with extreme living conditions and pronounced specificity of socio-economic development. And on the other hand, this macro-region is the zone of Russia's strategic interests due to unique geo-political, natural resources and socio-economic potential.

4 Conclusions

It is quite obvious, that in future Arctic will come up as the major resource of the oil and gas for the whole world. That is why the interest of the Arctic Countries in the natural resources of the Arctic will grow up and experts predict the struggle between them for those resources will become more acute.

Arctic shelf development is a prerequisite for economic power of the country and here, in addition to involvement in the process of extraction of oil and gas companies, an important factor in optimizing production processes and reducing related production risks is competent and well thought-out the state program for oil and other natural resources, and it is necessary to create conditions that stimulate the investment of extracting companies and guarantees for the protection of the funds invested in exploration.

The Arctic shelf is important for energy leadership of Russia. The Russian Federation has all advantages needed for the research and unimpeded movement over the Arctic region.

National Project of the Russian Federation provides the capability of oil reception from other fields. This will allow effective cooperation with the development of adjacent fields without the construction of similar platforms, making it cost-effective, due to the lower specific cost. Maximum output of the different platform will be reached by 2021, when oil production reaches maximum, and the cost per barrel will be about US$ 10 Gazprom does not cast doubt on the profitability of the project.

The creation of a new legal framework to manage the Arctic shelf industry in such a way as to guarantee access while protecting against unregulated exploitation seems to be the most urgent and complex task in considering uncertainties in the field of offshore exploration in the Arctic shelf area.

To make this step, new legal, political, operational and institutional frameworks will be necessary for the exploration and production industry to be able to address local, national and global concerns around Arctic offshore development. This step will require the active participation and involvement of a number of distinct stakeholders in the institutional framework.

References

1. Arruda, G.M.: Arctic governance regime: the last frontier for hydrocarbons exploitation. Int. J. Law Manag. **57**(5), 498–521 (2015). http://0-search.proquest.com.www.elgar.govt.nz/docview/1710151119?accountid=40858
2. Bamford, J.: Frozen assets: the newest front in global espionage is one of the least habitable locales on Earth—the Arctic. http://foreignpolicy.com/2015/05/11/frozen-assets-arctic-espionage-spying-new-cold-war-russia-canada/
3. Brigham, L.: Future perspective: the maritime Arctic in 2050. Fletcher Forum World Aff. **39** (1), 109–120 (2015). http://0-search.proquest.com.www.elgar.govt.nz/docview/1682906687?accountid=40858
4. Dmitrievsky, A.N., Belonin, M.D.: Prospects of development of oil and gas resources of the Russian shelf. Nature **9** (2014)
5. Federal law of Russian Federation No. 58-FZ: On procedures of foreign investments in economic companies which are of strategic significance for national defence and state security, 29 April 2008
6. Frolov, I.: Development of the Russian Arctic zone: challenges facing the renovation of transport and military infrastructure. Stud. Russ. Econ. Dev. **26**(6), 561–566 (2015)
7. Kozmenko: Geo-economic processes in Arctic and development of shipping routes. Publishing House of Kola Scientific Centre of RAS (Russian Academy of Sciences), Apatite (2014)
8. Official sites of the Gazprom and Rosneft. http://www.gazprom.ru, http://www.rosneft.ru. Accessed 23 Mar 2017
9. The Intergovernmental Panel on Climate Change Fifth Assessment Report (AR5). www.ipcc.ch/report/ar5; the Arctic Council's Arctic Climate Impact Assessment (ACIA). www.amap.no/arctic-climate-impact-assessment-acia; the Arctic Council's Arctic Marine Shipping Assessment (AMSA). www.pame.is/index.php/project/arctic-marine-shipping/amsa; and, the European Project Arctic Climate Change, Economy and Society (ACCESS). www.access-eu.org

Information Technology in Public Administration

Digital View on the Financial Statements' Consolidation in Russian Public Sector

Tatiana Antipova[(⊠)]

Federal State Budgetary Educational Institution of Higher Education
"Perm State Agro-Technological University Named After Academician D.N.
Pryanishnikov", Institute of Certified Specialists, Perm, Russia
antipovatatianav@gmail.com

Abstract. The major New Public Management reform in Russia began in 2004 and envisaged ideas of "top-down" modernization through adoption in national regulations of international reporting ideas from Government Financial Statistic. Based on online financial 2010–2014 public sector annual reports set on official websites open access this study analyzed the main elements from 2010–2014 Russian Annual questionnaire reports represented to International Monetary Fund in compliance with Government Finance Statistics Manual for the Integrated Correspondence System. The General government's consolidated financial statements consist of three sectors: Central, State and Local Governments. Herewith Central government sector includes such Social Security Funds as Pension Fund of the RF, Russian Social Insurance Fund, and Federal Compulsory Medical Insurance Fund that is different from Government Finance Statistics Manual comprehending. The uncertainty of Public Sector definition leads to the incomparability of Russian Public Sector Annual questionnaire reports. The research analyzes the main challenges from annual reports' consolidation in Russian public sector as a whole. The paper shows a scheme of the consolidation's process. The research result allows the transformation of reform ideas of into the "local" reporting practice action space characterized by two different accounting contexts: shaped by both "non-local" international reporting standards and the "local" Russian reporting tradition.

Keywords: Financial statements' consolidation · Annual reporting
Financial statements' legislation · Interbudgetary ties · Public sector
State Financial Report

1 Introduction

Consolidated Reporting of Government finance has recently become a very popular subject, with most of the discussions being focused on regulatory perspectives and reporting performance improvement. Matis and Cîrstea [11] point that the public sector consolidated financial statements might be considered a topic of current interest that is gradually developing. This issue of consolidated financial statements in the public sector has been of major interest lately even for researchers from different academic areas. As the topic of consolidated financial statements in the public sector is going to be implemented in different countries and it is under the process of developing in other

© Springer International Publishing AG, part of Springer Nature 2018
T. Antipova and Á. Rocha (Eds.): MosITS 2017, AISC 724, pp. 125–136, 2018.
https://doi.org/10.1007/978-3-319-74980-8_12

countries, from the information we have now there is no study that might have analyzed the specialized literature in the field of consolidated financial statements in the
public sector [11].

From the brief study of the prior literature, we can observe that Consolidated
Reporting has evolved to an independent concept by recognized international standard
bodies, especially International Monetary Fund (IMF) and International Public Sector
Accounting Standard Board (IPSASB). Financial statements of an entire country presented to the International Monetary Fund (IMF) are prepared and approved by the
member country's Ministry of Finance in accordance with Government Financial
Statistics Manual (GFSM 2014). All 188 IMF member countries (including Russia)
must comply with GFSM 2014.

Guidelines for responding to the annual GFS questionnaire provides instructions for
completing the Annual Government Finance Statistics (GFS) Questionnaire package,
which is comprised of Statistical Tables—that follow the framework of the IMF's
Government Finance Statistics Manual 2014 (GFSM 2014) methodology—and a
Metadata Questionnaire (formerly the Institutional Table). The IMF's Statistics
Department will present the information compiled and reported by countries in summary form in the Government Finance Statistics Yearbook (GFSY) and web-based
formats available on-line at www.imf.org. Research shows that even with the adoption
of GFS by many countries, the lack of accounting homogeneity is still evident between
adopting countries, meaning that variation of accounting practices is still an essential
characteristic of governmental financial reporting [2, 5, 9–13].

The major change in accounting came as a consequence of the Russian Federation
(hereafter – RF) establishing contact with IMF (E.g. see the decision of the Government of The RF from 22.05.1992 № 2815-1 «About the introduction of the RF into The
International Monetary Fund (IMF), International Bank for Reconstruction and
Development (IBRD) and International Development Association»). Since then, discussions went on about how financial statements' consolidations in Russia could be
transformed in accordance with IMF requirements. The major changes occurred in the
2004–2005 fiscal year as a consequence of Russia adopting the Concept [3]. This
resulted in new budgeting and accounting procedures in that registers were to be
introduced based on Government Financial Statistic classifications in accordance with
the Government Finance Statistics Manual [6]. The use of GFSM to change government budgeting and accounting were argued to be a major step forward in the compilation and presentation of fiscal statistics to improve government reporting and
transparency in its operations. Government finance statistics was argued to be a key to
fiscal analysis, reporting information was argued to play a vital role both in developing
and monitoring financial programmes and in conducting surveillance of economic
policies. Based on GFSM requirements, RF has introduced the accrual accounting
principle, including preparation of the Balance Sheet, and detailed coverage of government economic and financial activities. One way to see why this was important to
prepare these statements was legitimization in relation to IMF requirements which
asked for consolidated information for the whole government of RF. The main challenges of the financial statements' consolidation in Russia more or less corresponds to
the GFS system shown below. Accounting policies and the layout of accounting
financial statements are traditionally regulated by different instructions in Russia.

Current public sector financial accounting statements are regulated by the "Instructions on the order of compilation and presentation of the annual, quarterly, and monthly report on the execution Public Sector of the RF" approved by order № 191n and 33n of the Ministry of Finances of The RF (28.12.2010, 25.03.2011).

2 Methodology and Findings

Inside any country financial (including budgetary) statements of each separate public sector (budgetary) entity are prepared by accountants and approved by the head of this organization. Many regulations (e.g. Budget Code) and rules are usually ordered by the Ministry of Finance and spread inside the country. The financial statements of each budgetary entity presented depend on the level of budget system of this country. For instance, federal level budgetary entity presented their financial statements to the federal authority, etc. Many countries use IPSAS to accomplish this. Subsequently, financial statements are consolidated on each level of the budget system and then for whole country.

For whole country the Treasury of RF, in coordination with the Ministry of Finance, annually prepares the Financial Report of RF as State as a whole, hereafter referred to as the State Financial Report (SFR). The SFR is a general-purpose report of accountability intended internally for members of Government, state executives and program managers, top civil servants, and externally primarily for each citizens and intermediaries who are interested in and have a reasonable understanding of state activities. Citizen intermediaries include members of the news media, scientists, analysts, and others who study, analyze and interpret for the general public the more complex and detailed information in the SFR. The important goal of the SFR is to make available to every Russian a comprehensive overview of the state finances. But manly SFR is made for representation to IMF since RF is a member of IMF.

SFR is made according to the Instructions for Completing GFS Yearbook Questionnaire Report Forms for the Integrated Correspondence System (ICS) using consolidation method. Consolidation is a method of presenting statistics for a set of units as if they constituted a single unit. In the GFS system, the data for all sectors (central government sector, general government sector, nonfinancial public sector, public sector) should be presented on a consolidated basis. Consolidation involves the elimination of all significant transactions and debtor-creditor relationships that occur among the units being consolidated (units within a given subsector or sector) [7, p. 57, p. 22]. SFR consist of the required forms listed in Table 1.

Listed in Table 1 forms are necessary to fill out and represent each country IMF-member including Russia. Additionally they represent Statistical Tables and Annex. To do this Federal treasury of Russian Federation collects data from all of institutional units marked in GFS Yearbook Questionnaire Report Forms.

The reporting of separately taken level of budgetary system is consolidated by addition of indicators of organizes the execution of the budget, key administrators of budgetary funds, administrators of budgetary funds and budget users. If the reporting of

Table 1. The required forms for annually reporting State Financial Report to IMF.

GFSM 2014
Statement of operations
Statement of other economic flows
Balance sheet
Statement of sources and uses of cash
Statement of total changes in net worth
Summary statement of explicit contingent liabilities and net implicit obligations for future social security benefits

one of budget users is made incorrectly it generates problems at reporting consolidation. The approximate scheme of consolidation to SFR at federal level of RF budgetary system shown in Fig. 1.

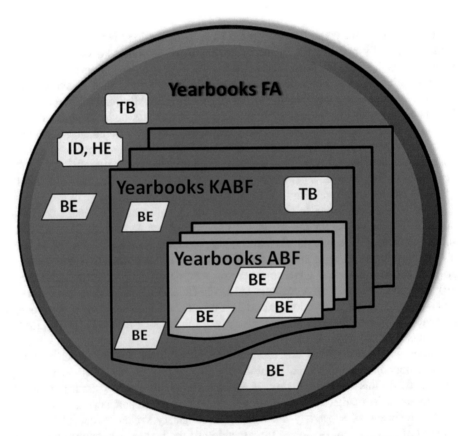

Fig. 1. The scheme of the budgetary reporting consolidation on federal level.

On the Fig. 1 we used the following abbreviations:

FA – the fiscal authority that organizes the execution of the budget;
KABF – the key administrator of budgetary funds;
ABF – the administrator of budgetary funds;
BE – budgetary and autonomous entities, budget users;
TB – territorial government bodies;
ID, HE – isolated divisions and/or head establishments without formation of the legal person.

The consolidations' challenges can be dividing on two categories: methodological and technical. Some methodological problems consist of: *(1) The absence of reporting definitions and interpretations and (2) The uncertainty consolidated elements of SFR.*

(1) *The absence of reporting definitions and interpretations*

Systems and rules in themselves provide a vital framework but cannot guarantee success if those who implement them do not have the right set of skills and competencies to implement them well and to make the technical and practical judgments required. Financial reports are reliable as long as the data they are supposed to present holds good quality. Thus, being of a different nature, the transformation from cash to accrual accounting depends on the improvement of skills and commitment to making them work. Skills and competencies will equally include the understanding of the systems to be operated, the stages of reforms and/or other changes taking place as well as the knowledge and intelligence needed to apply them in practice. With the introduction of accrual accounting, the focus on reporting use of assets and economic results in the accounting period has increased. There were reasonable expectations towards the methodological difficulties the Russian public sector accounting profession would meet in the use of accrual accounting and how these challenges could have been dealt with. However, the change expected by the professional accounting community towards more principle-based accounting has not so far been realized.

SFR is prepared through transferring the basic terms and concepts of GFSM into the rules and instructions of the traditional budgetary accounting. On the one hand, changes meant the introduction of new concepts and widening scope of accounting registrations and measurement. For instance, for the first time in Russian government accounting practice, concepts such as "non-financial" and "non-produced" assets were used. Thus, the balance sheet primarily reports assets and liabilities that represent net spendable and available resources for these funds. In addition to this objective, there are also more practical incentives for public sector bodies to structure projects so that the assets (and the corresponding liability to finance these assets) are recorded on their balance sheet.

(2) *The uncertainty consolidated elements of SFR*

There is the big ambiguity with institutional units' definitions. Russian law does not contain such important terms as "public sector", central government, general government, etc. for preparing SFR. None of Russian legislations define what is public sector and from what it consist. In Russian Budget Code is given just the term "budget system" definition. We can suppose that the term "budget system" is similar to "public sector" abroad. But it is not exactly because according to GFSM 2014 public sector comprises General Government and Public Corporations. Government-owned enterprises, such as the central bank, post office, or railroad are often referred to Public Corporations. General Government usually consists of three levels: Central Government, State or Regional Government, and Local Government. Public Corporations are divided on Nonfinancial (e.g. post office) and Financial (e.g. Central Bank). But Russian budget system consists of three levels: federal, subjects of Russian Federation, local governments, and does not include Public Corporations.

Based on digital view to online financial 2010–2014 public sector annual reports set on official websites open access this study analyzed the main elements from 2010–2014 Russian Annual questionnaire reports represented to International Monetary Fund (IMF) in compliance with Government Finance Statistics Manual (GFSM) for the Integrated Correspondence System. The General government's consolidated financial statements consist of three sectors: Central, State and Local Governments. Herewith Central government sector includes such Social Security Funds as Pension Fund of the RF, Russian Social Insurance Fund, and Federal Compulsory Medical Insurance Fund that is different from the GFSM comprehending. The uncertainty of Public Sector definition leads to the incomparability of Russian Public Sector Annual questionnaire reports. In addition inside each of institutional units we can see some variations that shown in Table 2.

Analysis Table 2 data shows that there is unclearly the reason including into Extrabudgetary units only two Government Corporation of seven functioned from 2007 to 2014. In addition, Social security fund data has not included in central government unit since 2014 fiscal year. Also we have to bear in mind that municipalities have never included in Russian general government due to the Russian Constitution. According to Art. 12 of the Russian Constitution, Local governments are independently and Local governments are not included in the system of public authorities therefore in public sector. Thus, when total sum revenue and expenses counted Local government indicators not included. We can see it from data shown in Table 2. For data comparability social funds are included in the central government for 2014 fiscal year. Also there are unknown how many Institutions of local governments were considered in 2009 – 2011. Using digital view author was analyzed Russian Federation annual GFS questionnaires. Result of this analysis is shown in Table 3.

The analysis of Table 3 allows suggesting that General Government sector in total should consist of the data of the central government, regions and municipalities. But in fact, the rules of arithmetic do not work here, and it is impossible to determine why the 2014 general government revenues are equal to 31045,5 billion rubles since it is not the sum 15757,3 + 11049,8 + 6065,2 = 32872,3. IMF Statistical Tables contain Consolidation Columns minus some of the data due to unknown reason, making it impossible to understand the origin of the totals. In our opinion this is another methodological problem.

Table 2. The institutional units of Russian General government for SFR.

Fiscal year	General government				
	Central government			Regional governments	Local governments
	Budgetary	Extrabudgetary	Social security funds		
2014	Federal government agencies	Federal budgetary and autonomous entities + the Government corporations of RF: "Fund of assistance to housing and communal services reforming" and "Russian Corporation of Nanotechnologies" + the state company "Russian Roads"	Mandatory medical insurance funds – federal and 85 territorial; Pension fund of the RF; RF social insurance fund	85 subjects of RF	22750 municipalities
2013	Federal government agencies	Federal budgetary and autonomous entities + the Government corporations of RF: "Fund of assistance to housing and communal services reforming" and "Russian Corporation of Nanotechnologies" + the state company "Russian Roads"	Mandatory medical insurance funds – federal and 83 territorial; Pension fund of RF; RF social insurance fund	83 subjects of RF	23133 municipalities
2012	Federal government agencies	Federal budgetary and autonomous entities + the Government corporations of RF: "Fund of assistance to housing and communal services reforming", and "Russian Corporation of Nanotechnologies" + the state company "Russian Roads"	Mandatory medical insurance funds – federal and 83 territorial; Pension fund of the RF; RF social insurance fund	83 subjects of RF	23102 municipalities

(continued)

Table 2. (*continued*)

Fiscal year	General government				
	Central government			Regional governments	Local governments
	Budgetary	Extrabudgetary	Social security funds		
2011	Federal government agencies	Federal budgetary and autonomous entities + the Government corporations of RF: "Fund of assistance to housing and communal services reforming" and "Russian Corporation of Nanotechnologies" + the state company "Russian Roads"	Mandatory medical insurance funds – federal and 83 territorial; Pension fund of the RF; RF social insurance fund	83 subjects of RF	Institutions of local governments
2010	Federal institutions of the government of the RF	Extrabudgetary resources of the federal budgetary institutions + the Government corporations of RF: "Fund of assistance to housing and communal services reforming", and "Russian Corporation of Nanotechnologies" + government company "Russian highways"	Mandatory medical insurance funds – federal and 83 territorial; Pension fund of the RF; RF social insurance fund	83 subjects of RF	Institutions of local governments

Source: Author's analysis of data retrieved from http://roskazna.ru/ispolnenie-byudzhetov/statistika-gosudarstvennykh-finansov-rf/

Table 3. The Russian Federation annual GFS questionnaires analysis. In Billions of Rubles/Fiscal Year Ends December 31 government for SFR

Institutional units	Fiscal year	Total revenue	Taxes	Social contributions	Grants	Other revenue
Central government	2010	12365,3	6042,5	2508,8	300,0	3514,0
	2011	17911,5	8388,4	3493,5	389,3	5640,3
	2012	19001,3	9367,4	3993,0	451,5	5189,4
	2013	19579,3	9439,6	4932,6	207,8	4999,3
	2014	15757,3	10475,7	1,3	75,5	5204,9
State governments (subjects of Federation)	2010	6273,1	3734,8	0,0	1568,5	969,8
	2011	7510,0	4220,3	0,0	1874,1	1415,6
	2012	10233,3	4893,8	0,0	3309,0	2030,5
	2013	10408,8	4857,3	0,0	3363,8	2187,7
	2014	11049,8	5316,1	0,0	3458,9	2274,8
Local governments	2010	3750,3	773,1	0,0	1649,2	1328,0
	2011	4247,4	801,5	0,0	2004,6	1441,3
	2012	5412,6	922,7	0,0	3077,0	1412,9
	2013	5979,9	1027,0	0,0	3300,7	1652,2
	2014	6065,2	950,7	0,0	3653,1	1461,4
General government	2010	18864,9	10550,4	2508,8	0,6	5805,1
	2011	25389,7	13410,2	3493,5	0,0	8486,0
	2012	27796,6	15183,9	3993,0	0,0	8619,7
	2013	28752,2	15323,9	4601,9	0,0	8826,4
	2014	31045,5	16742,6	5222,0	0,1	9080,8

In addition technical challenges may be classified as:

- *The increasing complexity,*
- *A lack of accounting knowledge;*
- *The information technology advances.*
- *The increasing complexity.* New public sector accounting has become much more complex and formalized. For example, earlier versions of the Chart of Accounts consisted of 120 accounts and each account was marked with a 3 digit code for registration. Based on Public sector accounting instructions, the budgetary accounting should be conducted according to the Chart of Accounts consisting of more than 1583 analytical accounts and reflect classification in accordance with the Budget Code in order to trace actual performance compared to the budgetary allocations.

Public sector accountants experienced problems explaining the benefits of the increasingly complex accounting system. It seems that the reasons and necessities for implementing reforms were not clearly communicated by reformers to the accounting profession. For instance, instructions did not communicate clearly the tasks of accrual accounting and its potential areas of usage. Thus, it seems that the public sector accounting profession has received a new accounting tool, which is claimed to be better suited for providing accounting reports of better quality, but nobody has told the accountants why these accounting tools were needed and how to use them properly.

- *A lack of accounting knowledge.* Improvements in competencies will require regular training well suited to the requirements of the specific work to be done. Professional knowledge is thus an important indication of how well the professional accountant is interested in and prepared to contribute to accounting reforms.

In this respect, it is interesting to notice the reaction of the Russian accounting profession to the introduction of new accrual accounting rules. Reforms are based on the adoption of Western accounting norms and regulations, both GFS and IPSASs. We have asked for both the degree of GFSM and IPSAS knowledge [4]. It is striking that the majority of accountants are not even aware either of GFSM or of IPSAS and seem unfamiliar with these documents.

For the accuracy and reliability of data each of report preparers must understand the statements elements identically. But due to the absence of education system for public sector accountants it is very difficult in Russia. Only a few institutions in the whole of Russia offer special graduate education programs and/or competence improvement programs in public sector accounting (e.g. in Moscow just some State Universities provide this - "Financial university", "Moscow University of Ministry of Internal Affairs of Russia" and others). Therefore, most public sector accountants seemed to be employed with a general economic education. Most training in public sector accounting is carried out using the "master-apprentice" model, i.e. at work, by which a more experienced accountant transfers his/her knowledge to less experienced employees. It can be expected that such training reinforces the instructional and compliance oriented accounting tradition.

- **The information technology advances.** Given the new challenges related to environmental changes and technological advances in recent years, clarification for implementation and sustainment becomes critical. In addition to these challenges, many new software-development techniques and architectures such as cloud service, shared service, agile development and spiral development, which focus on module-based swift development with much shorter development cycles, pose new issues [1]. Reporting measurement to citizens is essential to help them ascertain related to such issues as equity and citizen satisfaction. The future of Russian governmental accounting will be determined largely by the ways in which conflicts between two contrasting viewpoints will unfold and be resolved.

3   Conclusions and Discussions

This paper analyzed just some challenges of financial statements' consolidation in Russian Public Sector with digital view. The research result allows the transformation of reform ideas of into the "local" reporting practice action space characterized by two different accounting contexts: shaped by both "non-local" international reporting standards and the "local" Russian reporting tradition.

In considering how to further improve the public sector financial reporting process in their countries, researchers thought that the following measures were a high priority [8, p. 21]:

- Continue convergence to one global set of financial reporting standards
- Globally unify, simplify, and clarify financial reporting standards focusing more on the best practice
- Ensure that boards of directors pay attention to the quality of financial reports
- Provide additional education and training for preparers.

We can discuss with Allen [1] a couple of broad areas for improvement. First, financial reporting for the whole country needs to reach a level of maturity such that it is looked to as an anchor of integrity despite how political winds may blow. Secondly, financial statements need to be restructured so there is a bottom line to the financial statements that reflects in a clear, understandable manner the improvement or deterioration of the financial condition of the general government each reporting period. A lot of general financial information is reported, but even financial experts differ on whether the financial condition of the federal government is improving or deteriorating [1].

While recognizing these challenges, it is considered important that the issues involved are not just either ignored or left for each country to resolve for itself. To do so seriously undermines the ability of countries to reform and dilutes the impact of any technical assistance provided.

Demonstrating accountability for resources, obligations and financial affairs is among the stated objectives of public sector financial reporting. Producing reliable financial statements should be viewed as a byproduct of effective business processes and financial management systems. The primary goal is to improve financial management systems so that financial information from these systems can be used to help manage agencies more effectively. Well-designed and operating integrated financial management systems significantly reduce the cost of preparing financial statements. In addition, agencies would have the reliable, useful and timely information from such systems to assist them in managing their operations day-to-day.

There is a need for a better collaboration between accounting academia and accounting practitioners. There are successful examples of how reform ideas can be materialized in practices by researchers taking on consultants' role in developing accounting innovations in the public sector. This can be done by several means in Russia. For instance, representatives from accounting professionals and academia can be actively involved in a revision and simplification of new accounting standards. Academia can also be involved in more action oriented research at the level of particular organizations to develop (and later, to discuss and disseminate) best practices of accrual accounting adaptation. By doing so, it is possible to bring together the old practices and new ideas in atmosphere of debate and professional dialog.

When users are confident that the information they receive is relevant, reliable, understandable, consistent and comparable, this creates trust. Transparency and public accountability further engender trust in a representative democracy. Working together, these factors lead to greater citizen satisfaction and better access to capital at a lower cost.

References

1. Allen, T.L.: Reporting improved but challenges remain. J. Gov. Financ. Manag. **64**(3), 10–11 (2015)
2. Antipova, T.: Auditing for financial reporting. In: Global Encyclopedia of Public Administration, Public Policy, and Governance. Springer (2016). https://doi.org/10.1007/978-3-319-31816-5_2304-1
3. Antipova, T.: Audit of budgetary entities. Novosibirsk, in Russian (2008)
4. Antipova, T., Bourmistrov, A.: Is Russian public sector accounting in the process of modernization? An analysis of accounting reforms in Russia. Financ. Account. Manag. **29**(4), 442–478 (2013)
5. Benito, B., Brusca, I., Montesinos, V.: The harmonization of government financial information systems: the role of the IPSASs. Int. Rev. Admin. Sci. **73**(2), 293–317 (2007)
6. Government finance statistics manual: International Monetary Fund, Washington, DC (2014)
7. Government Finance Statistics: Compilation Guide for Developing Countries: International Monetary Fund, Washington, DC (2011)
8. International Federation of Accountants (IFAC): Financial reporting supply chain. Current perspectives and directions (2008)
9. Kristofík, P., Lament, M., Musa, H.: The reporting of non-financial information and the rationale for its standardisation. E+M Ekon. Manag. **19**(2), 157–175 (2016)
10. James, M.L.: Accounting majors' perceptions of the advantages and disadvantages of sustainability and integrated reporting. J. Legal Ethical Regul. Issues **18**(2), 107–123 (2015)
11. Matis, D., Cîrstea, A.: Reflections on public sector consolidated financial statements research. Stud. Univ. Babes-Bolyai **60**(3), 69–82 (2015)
12. Oprisor, T.: Contributions to an improved framework. Account. Manag. Inf. Syst. **14**(3), 483–507 (2015)
13. Warren, K.: Time to look again at accrual budgeting. OECD J. Budg. **14**(3), 113–129 (2015)

How to Be Open About Spending: Innovating in Public Sector Reporting in the Information Age

A. S. C. Faber[(⊠)] and G. T. Budding

School of Business and Economics, Vrije Universiteit Amsterdam,
De Boelelaan 1105, room 6A-59, 1081 HV Amsterdam, The Netherlands
a.s.c.faber@vu.nl

Abstract. The field of public sector reporting is moving from vertical to horizontal accountability: from hierarchical relationships that mediate between public organisations and citizens to directly disclosing available information. IT developments play an important role in innovative forms of reporting that respond to the growing importance of horizontal accountability. This paper provides an overview of some of the most important initiatives on the plane of reporting innovations, such as *gov.uk* and *OpenSpending*, as well as recent examples in the Netherlands. The examples underline the importance of horizontal accountability forms, as well as the significance of 'horizontal integration': the idea that systems are fully integrated across different functions. However, issues of interoperability, security, and information architecture persist, often pared with the lack of authority to coerce (cross-sectoral) reporting changes. The paper concludes with a discussion of how these initiatives can aid further improvement in public accountability. We argue that the new forms of reporting do not only influence the question *how* reporting is composed, but that they also affect *wha*t is reported and for *whom* the reporting is compiled.

Keywords: Public accountability · Public sector reporting · ICT
E-government · Interoperability

1 Introduction

Constant changes in society demand an adaptive approach in all areas of the public sector. One of the areas on which changing societal conditions exert their influence on the public sector is the field of accountability. Traditionally, accountability is seen as a relationship that consists of giving and demanding reasons for conduct [1]. Over time, a range of mechanisms was developed in order to live up to this basic view of accountability, such as public sector reporting. However, these mechanisms have demanded an update with the aim of adapting to the societal changes and to better meet the information needs of users. Innovating in the post-*New Public Management* era [2] transcends the fine-tuning of existing structures and is much about the utilisation of new technical possibilities that help to fulfil the high expectations of citizens in the information age [3]. Therefore, innovation in public accountability goes beyond small reforms of the budget and annual report.

© Springer International Publishing AG, part of Springer Nature 2018
T. Antipova and Á. Rocha (Eds.): MosITS 2017, AISC 724, pp. 137–147, 2018.
https://doi.org/10.1007/978-3-319-74980-8_13

In the field of public sector reporting, the accent on vertical accountability is slowly shifting to horizontal accountability: a transition is made from hierarchical relationships that mediate between public organisations and citizens to directly disclosing the available information. This paper provides an exploratory overview of a selection of initiatives in reporting innovations by means of the interaction between government and citizens, standardisation, and data portals. The paper concludes with a discussion of what consequences these initiatives can have for further improvement in public accountability.

2   Historical Streams: From Vertical to Horizontal Accountability, from Internal to External User

Traditionally, most OECD countries are characterised by a vertical accountability structure [4]. Vertical accountability of governmental organisations is based on the powers and responsibilities of the actors in these organisations, and its content is linked to the exercise of their powers and the fulfilment of their responsibilities [5]. Politically, public managers are accountable only *organisationally*: basically they are only responsible to their immediate superior within their column. Only the top of this organisational pyramid – the minister – can be held accountable by parliament, which in turn is the mediator for citizens and the media. In this way, vertical public accountability is controlled by a chain of principal-agents (see Fig. 1).

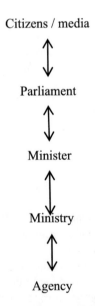

Fig. 1. Vertical accountability (adapted from [4])

In the last decades, the accent on vertical, hierarchical forms of public account-ability is being refocused (One of the reasons for the changes in accountability was the rise of so-called 'administrative' accountability relationships through agencies such as auditors and independent inspections. Their reports are disclosed to parliament and ministers, but the administrative accountability relationships cannot exert direct influ-ence. Therefore this is considered *diagonal* accountability: these agencies are meant to promote parliamentary control, but they are not part of the direct principal-agent-relationship. For our argument, horizontal accountability is the more relevant development. See e.g. [6]). As the 1990s progressed, the focus in public sector developments turned to the growing importance of values such as 'trust' and 'trans-parency' [7]. With this, a growth took place of the importance of a direct and explicit accountability relationship of government with citizens, civil society organisations (CSOs), and the media, as they can be considered the eventual stakeholders at the end of the chain. This commonly referred to as horizontal accountability (See Fig. 2). The historical movement from vertical to more horizontal forms of accountability is also confirmed by current literature (See e.g. [8]. On the one hand, actors are now more directly accountable to citizens, media, and CSOs. On the other hand, the Ombudsman, auditors, and supervisors can now also move more freely in exercising their supervi-sory functions. The principal-agent chain of accountability thus persists, but is sup-plemented and amplified by horizontal forms.

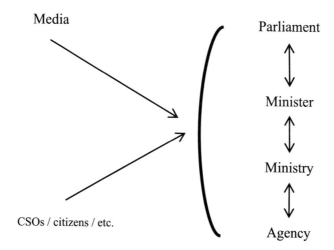

Fig. 2. The combination of vertical accountability with horizontal accountability (adapted from [4])

Next to the difference between vertical and horizontal accountability, a distinction can be made between internal and external users of information. As the distinction between internal users (e.g., management) and external users (e.g., shareholders) in the private sector is rather clear, this is more complicated in the public sector [9]. In the public sector internal reporting is more often confronted with external regulation and requirements for detailed financial information. Reasons for this are that the stakeholder is not only relatively broader (every taxpaying citizen or organisation can be defined as a stakeholder (For definitions of taxpayers as stakeholders, see e.g. [10])), but also better equipped to exert influence (See e.g. [11]).

Furthermore, the private sector shows an explicit disconnect between internal and external reporting: there are documents explicitly meant for internal use such as management reports and dashboards, whereas the annual report is for external use. In the case of the public sector, the functions of internal and external reporting tend to coincide much more. This stimulates the growing importance of horizontal account-ability. The growing coincidence of internal and external information needs causes (unprocessed, raw) data to also find its way towards external stakeholders. This movement is accelerated because of technologies which can be directly linked to particular datasets. Therefore it could even be argued that there is no need to facilitate a particular 'shell' for this information, in the form of reporting – either internal or external.

It could be argued that with the emergence of open data, the external stakeholder in the public sector context (the citizen, the media) is relieved from its previous role of *end-user* in the accountability chain: she or he does not receive 'ready-for-use' accounting information. Instead it is supposed that the user starts to work with, and give meaning to the numbers. The expectations associated with the disclosure of open data are that, in the long run, citizens, companies and institutions actively reflect on the data, and look over the shoulder of policy makers and fulfil their role as so-called 'armchair auditors'. However, current literature is still critical of the – up to date often scarce – degree to which actors take up the intended role of armchair auditor (See e.g. [12]). Actual working with the data by parties or individuals outside government does not happen structurally; often it happens only then when the government itself facili-tates this, for example in the form of hackathons. In the Netherlands, hackathons and hacker meet-ups are organised frequently in the public sector, ranging from events under the auspices of the Supreme Audit Institution and the Ministry of Internal Affairs to the Dutch Royal Meteorological Institute.

3   Innovations in Public Sector Reporting

In this paragraph we look at how organisations deal with accountability in the infor-mation age by means of innovating in public sector reporting, making use of infor-mation technology. The examples highlighted were chosen based on scale, level of complexity and the playing field in which the organisations operate. They furthermore form an illustration for public sector institutions that have to act in a complex, multi-stakeholder playing field.

gov.uk, set up the United Kingdom, is commonly considered one of the forerunners when it comes to innovative uses of information technology in public accountability. The project of *gov.uk* is led by the Government Digital Service. The website was to replace the individual websites of hundreds of government departments and public bodies by 2014. Up to date, these are to a very large degree aggregated and merged into this portal [13]. The main page of the website integrates service delivery-oriented content with other (financial as well as nonfinancial) government information. *gov.uk* is set up as a portal, with most prominently featured a search bar and a 'Services and information' section, consisting of an assortment of categories such as 'taxes', 'benefits' and 'jobs'. The next section on the main page is titled 'Departments and policy' and provides access to 'all policies, announcements, publications, statistics and consultations', in addition to a page providing an overview of available performance dashboards. The final item before the generic website footer consists of an elementary section called 'Help us improve *gov.uk*' with two to-be-filled-in open fields: 'What you were doing' and 'What went wrong'. Accompanying *gov.uk* was the launch of *data.gov.uk*: a public data catalogue pointing to around 20,000 datasets downloadable under open government license [14]. Moreover, *gov.uk* includes a range of open source-projects, in which the public is invited to think along about solutions to possible issues [15]. Although the website is generally very well received and appreciated, (See e.g. [16]) it was underlined that the creation of such a website is not simple, and complex issues of integration and interoperability, security, and information architecture must to be taken into consideration [17].

An example in the context of development work is the ***International Aid Transparency Initiative** (IATI)*. The exchange of aid data between governments and systems has often been characterised as a highly manual, labour-intensive process. Because of the wide range of institutions that organisations working in development are accountable to, which in turn have different information needs, time and again the accountability to every separate stakeholder has to be compiled and tailored. The *IATI* aims to build a common standard fostering consistency in how data on development cooperation activities is reported [18]. This data is intended for use by all organisations engaged in development, ranging from government donors to private sector organisations, as well as (inter-)national NGOs: thus the addressee of the information submitted to the IATI is made very explicit. Data that deploys the standard can also be submitted by all of these organisations. Its ultimate goal is to standardise data, so that information systems can automatically share relevant information, thus enabling improved decision making, greater accountability, results-based management, and comparative performance evaluations across countries and donors [19]. In addition, the data have to be timely, comprehensive, forward-looking, structured, comparable, and open. There are two options to publish an IATI standard:

1. *Organisation Standard*, consisting of forward-looking budgets, strategic documents such as country plans and annual reports, and budgets per country or region.
2. *Activity Standard*, consisting of full transaction history, sub-national geographic coding on the location, sectors and classifications, forward-looking budgets per activity, and conditions that are attached to activities, and results.

The 'ecosystem' of the IATI, aggregating all of the submitted data, works to provide a system through which IATI data can be accessed, tracked and queried. The adage to this approach has been '*publish once, use often*': [20] data have to be published only once, after which every stakeholder can extract the items and information that is most relevant to his information needs. This would in turn lead to a great relieve in administrative burden for affiliated organisations.

At the time of writing, 587 organisations have published data through the IATI portal [21]. Unfortunately, the aggregation of data is still difficult due to the reporting traditions between stakeholders that still vary widely. This shows that the amount of authority of the organisation charged with the lead function can be decisive for the success of integration and standardisation; in the case of *gov.uk*, the Government Digital Service was in the position to coerce organisations to the supply data in a uniform, standardised format nationwide. As a multi-stakeholder institution, the IATI can provide the conditions under which a standardised and comprehensive dataset can be realised, but due to the vast playing field of development cooperation, it can only to a very limited degree enforce that the relevant parties commit to providing their information in the form of the standard.

Finally, *OpenSpending* is a largely crowdsourced project that aims to track and analyse public spending worldwide. At the time of writing, the site hosts 2,266 data packages from 77 countries. Datasets can be submitted and modified by anyone, supplemented by a 'sanity check' from its operators [22]. It hosts transactional and budgetary data, with a focus on government finance. It contains the records in a structured form stored in database tables and provides searching and filtering as well as visualisations. The audience for OpenSpending is addressed differently at various instances: on one place it is grouped in 'contributors' or 'publishers', 'developers' or 'hackers', and 'civil society' or 'users'; [23] Elsewhere the user of the platform is defined as 'any interested stakeholder' [24].

Governments underlined that it sees initiatives like *OpenSpending* as the solution to the problem of interpreting public sector information [25]. However, some caveats have to be named. As with the IATI, the initiative is largely dependent on the beneficence of the contributors of datasets. This causes large, almost insuperable discrepancies in availability of data, with for example 44 datasets for the Caribbean isle of Saint Lucia against only 13 datasets for the entire African continent [26]. Moreover, the datasets themselves vary greatly in quality and level of detail, which also raises questions if a standard is applied. However, there are more than 30 local sites that are (partially) powered by *OpenSpending* architecture, which make attempts in detailing the available information in more elaborate ways than is the case with the overarching initiative.

3.1   Dutch Examples of Innovations in Public Sector Reporting

For the next part, we will discuss three Dutch innovations on the plane of public sector reporting: the Dutch branch of *OpenSpending*, the pilot application *Wat Doet De Poet?*, and the online representation of the annual report of the public utility company Alliander.

In the Netherlands *OpenSpending* is maintained by the non-profit Open State Foundation and is linked to the open data portal of Statistics Netherlands, the governmental institution gathering statistical information about the Netherlands [27]. Statistics Netherlands in fact already documented and disclosed a great deal of (financial) information on decentral bodies such as municipalities and provinces. Data stems from the so-called Iv3-standard, meaning Information for Third Parties. This information is provided every quarter of a year by local governments to Statistics Netherlands. A great benefit of this is that the data has already been standardised and (roughly) checked for errors and anomalies, thus facilitating in-depth research and comparison without any large prior transformations that have to be dealt with first.

A primary function of the website is the ability to compare the financial statements of different local governments. The application provides functionalities allowing comparisons on different ways and levels. As a result, the website provides a specific interpretation of the *OpenSpending* architecture: it functions primarily as a comparative and benchmarking tool. In 2014, a preliminary study was conducted after the use and the usability of *OpenSpending* in the context of the city of Amsterdam by three stakeholder groups: politicians, city officials, and journalists [28]. The degree of transparency achieved through the initiative was reviewed positively by the three groups. However, the stakeholders underlined that timeliness, more detailed information, and comprehensiveness of the datasets were a prerequisite in order for the platform to be used efficiently. Additionally, this shows that the benefits of information technology are not reserved for the external user, but that the internal user (in this case: the city official) can also put it to good use. The Open State Foundation asserts that OpenSpending is still under construction, and aims to further develop the level of detail in the data

A similar operationalisation is the pilot application **Wat Doet De Poet?** (*Wat Does The Money Do?* Hereafter: WDDP) [29]. In the Netherlands the central government determines to a large extent what tasks are executed by municipalities. However, municipalities are fairly free to decide how these tasks are organised, leading to greater differences between municipalities. This makes it easy to trace differences in spending patterns between municipalities by means of open data. Just as was the case with *OpenSpending*, this application gets its spending information from the Iv3-portal of Statistics Netherlands, as well as a selection of performance indicators that are available as open data from a comprehensive municipal benchmark database maintained by the Kwaliteitsinstituut Nederlandse Gemeenten (*Quality Institute for Dutch Municipalities*) [30]. *WDDP* allows filtering by relevant municipality and particular field of interest such as education, with the aim of disclosing the information deemed most relevant by the user. After the input has been processed, a dashboard shows how much money has been spent in the particular area by means of relevant cost items, and it shows if the money is spent efficiently by means of performance indicators. With this, *WDDP* attempts to relate financial to nonfinancial information. In addition to *Wat Doet De Poet?*, the award-winning pilot-application developed in the same hackathon *De Gemeente Deler* is worth mentioning, which similarly makes an attempt to link financial to nonfinancial information.

The experiments and innovations regarding public sector reporting are not constrained to the macrolevel of (local) government. An example of a public organisation that experiments with IT innovations in its year report is **Alliander**, a public utility company that takes care of the distribution of energy in one-third of the households in the Netherlands [31]. The company distinguishes three main stakeholders as the addressee of its reporting: (1) employees, (2) customers, and (3) shareholders and investors. For its 2016 annual report, the three groups were asked which subjects they would like to be informed about in the report. Accordingly, fifteen top themes were selected ranging from 'safe and healthy work' to 'energy transition' and 'chain responsibility', forming the backbone of the annual report. These subjects were in turn grouped under the sustainability reporting guidelines formulated by the Global Reporting Initiative. Next to this, financial and nonfinancial information for Alliander and its subsidiaries is consolidated according to the methodology of Integrated Reporting. Finally, the report has an option to compile 'My Report', which allows the user to select the aspects relevant to his/her situation, which then gets compiled into a personally tailored annual report. The functionality to compile a personal annual report has been adapted by a range of other public sector organisations in the Netherlands as well; the case of Alliander serves as an illustration. Although this functionality appears as a way to oppose the risks of information overload, the actor looking for a report tailored to her or his situation still has to make a selection out all of the annual report's components, which in fact might amount to the same as scanning the annual report for relevant information.

4   Summary and Conclusions

Present-day developments in information technology offer new possibilities for public accountability through information disclosure. This paper provided a first selection of innovations in public sector reporting. The examples discussed underline the growing importance of horizontal accountability: all of the initiatives seek out for other forms of accountability that supplement (and, ideally, *com*plement) traditional forms of vertical accountability. These initiatives all use open data and provide a 'shell' so that the data is made accessible to a broader audience, thus creating opportunities for non-technical users to make sense of government data. By offering a choice to filter out the most relevant parts of accountability applying to her or his situation, innovations such as *OpenSpending* and *WDDP* foster the notion that the debate is not only focused around the question how better accountability can be reached, but also how a greater degree of *responsiveness* can be brought across to the public.

Furthermore the initiatives underline not only the addition of horizontal account-ability, but also the increasing significance of 'horizontal integration': the idea that systems can be fully integrated across different functions. *gov.uk*, in its integrative design that merges dimensions of service-delivery with public accountability, comes closest to the idea of government as a 'real one stop shop for citizens' [32]. However, other initiatives such as the *IATI* and the international *OpenSpending* show that stan-dardisation is still a major hurdle, and that it is a great investment in time, money, and the availability of the right power and authority to make the difference, and to make system not only workable and operative, but also comprehensive. In the case of *gov.uk*, the Government Digital Service was in the position to coerce organisations to the supply data in a uniform, standardised format nationwide. But the IATI, as a multi-stakeholder institution, can provide the conditions under which a standardised dataset can be realised, but it can only to a very limited degree enforce that the relevant parties commit to providing their information in the form of the standard. In all of the discussed initiatives, the accent seems to lie on finances, but innovations are also experimenting with how to incorporate performance data, in one case (*WDDP*) even taking first steps to connect this to financial data.

Finally, we argue that the new reporting forms do not only influence the question *how* public sector reporting is done, but that they also affect other questions. We contend that innovations are not limited to the representational dimension. They also have implications for the addressee (also referred to as *accountee* [33]) of the report, for *whom* the account is given: the data provider can for example become the most prominent user of an application (in the case of the Dutch *OpenSpending*), and the availability of new forms can evoke new users who were previously unable to work with the raw data.

Additionally, next to the question of how and for whom is accounted, innovations in public sector reporting also can have consequences for the subject of the report, on the question *what* is accounted. A 'One-Size-Fits-All'-approach does not hold when it comes to giving an account: a single database cannot be tailored to the needs of every potential user without the risk of information overload. Therefore, the innovations point towards the tendency to what can be called *tailored accountability*: accountability that is much more demand-driven than it is supply-driven [34]. However, the presentational layers in turn can decide what is presented, as the accountability becomes clickable and tailored to a personal situation (*WDDP* and Alliander).

References

1. Roberts, J., Scapens, R.: Accounting systems and systems of accountability: understanding accounting practices in their organisational context. Account. Organ. Soc. **10**(4), 443–456 (1985)
2. Pollitt, C.: Managerialism Redux? Keynote Address, EIASM Conference, Edinburgh (2014)
3. Castells, M.: The Rise of the Network Society. Wiley, New York (2011). Cordella, A., Bonina, C.M.: A public value perspective for ICT enabled public sector reforms: a theoretical reflection. Govern. Inform. Q. **29**, 512–520 (2012)
4. Bovens, M.: Public accountability. In: Ferlie, E., et al. (eds.) The Oxford Handbook of Public Management, pp. 182–208. Oxford University Press (2007)
5. Dees, M.: External reporting by public-sector organisations identifying new topics with a bearing on accountability. ECA J. **6**, 11–14 (2011)
6. Magnette, P.: Between parliamentary control and the rule of law: the political role of the Ombudsman in the European Union. J. Eur. Public Policy **10**(5), 677–694 (2003)
7. Pollitt, C., Bouckaert, G.: Public Management Reform: A Comparative Analysis New Public Management, and the Neo-Weberian state. University Press, Oxford (2011)
8. Willems, T., Van Dooren, W.: Coming to terms with accountability. Combining multiple forums and functions. Public Manage. Rev. **14**(7), 1011–1036 (2012). Klijn, E.H., Koppenjan, J.F.M.: Accountable networks. In: Bovens, M., Goodin, R.E., Schillemans, T. (eds.) The Oxford Handbook of Public Accountability, pp. 242–257. Oxford University Press (2014)
9. Schoute, M., Budding, G.T.: Stakeholders' information needs, cost system design, and cost system effectiveness in Dutch Local Government. Financ. Accountability Manage. **33**(1), 77–101 (2017)
10. IPSASB: The conceptual framework for general purpose financial reporting by public sector entities. IFAC Report, October (2014). US GASB: Why Governmental Accounting and Financial Reporting Is – and Should Be – Different. White Paper (2013)

11. Carvalho, J.B.D.C., Gomes, P.S., Fernandes, M.J.: The main determinants of the use of the cost accounting system in Portuguese Local Government. Financ. Accountability Manage. **28**(3), 306–334 (2012)

12. O'Leary, D.E., Armchair Auditors: Crowdsourcing analysis of government expenditures. J. Emerg. Technol. Account. **12**, 71–91 (2015)

13. GDS: gov.uk (2017). https://www.gov.uk/. Accessed 20 Oct 2017

14. Shadbolt, N., et al.: Linked open government data: lessons from Datagovuk. IEEE Comput. Soc. **27**(3), 16–24 (2012)

15. Government Digital Service: Git Hub – main page (2017). https://github.com/alphagov/. Accessed 20 Oct 2017

16. Wainwright, O.: Direct and well-mannered' government website named design of the year. The Guardian, 16 April 2013. https://www.theguardian.com/artanddesign/2013/apr/16/government-website-design-of-year. Accessed 20 Oct 2017. Author unknown: Writing for Design/Writing for Websites & Digital Design, D&AD (2013). http://www.dandad.org/awards/professional/2013/writing-for-design/20081/govuk/. Accessed 20 Oct 2017

17. Sarantis, D., Askounis, D.: Knowledge exploitation via ontology development in e-government project management. Int. J.Digital Soc. **1**(4), 246–255 (2010)

18. Development Initiatives: Implementing IATI: practical proposals (2010). http://www.aidtransparency.net/wp-content/uploads/2009/06/Implementing-IATIJan-2010-v2.pdf. Accessed 20 Oct 2017

19. Linders, D.: Towards open development: Leveraging open data to improve the planning and coordination of international aid. Govern. Inform. Q. **30**(4), 426–434 (2013)

20. Development Initiatives (2010)

21. International Aid Transparency Initiative: IATI Data (2017). http://iatistandard.org/202/guidance. Accessed 20 Oct 2017

22. OpenSpending Community: About OpenSpending (2017). http://community.openspending.org/about/. Accessed 20 Oct 2017

23. OpenSpending Team: OpenSpending platform update, Open Knowledge Blog (2017). https://blog.okfn.org/2017/08/16/openspending-platform-update/. Accessed 20 Oct 2017. OpenSpending Team: OpenSpending Documentation (2017). http://docs.openspending.org/en/latest/. Accessed 20 Oct 2017

24. OpenSpending Team: What is the Open Fiscal Data Package? Open Knowledge Blog (2016). https://blog.okfn.org/2016/10/20/what-is-the-open-fiscal-data-package/. Accessed 20 Oct 2017

25. Magure, S.: Can data deliver better government? Polit. Q. **82**(4), 522–525 (2011)

26. OpenSpending Team: OS Explorer (2017). https://openspending.org/s/?viewAll. Accessed 20 Oct 2017

27. Open State Foundation, Open Spending (2015). https://openstate.eu/nl/projecten/politieke-transparantie/open-spending/. Accessed 20 Oct 2017

28. Baars, D., et al.: Open Spending. Een kwalitatief verkennende dieptestudie naar de beoordeling en het gebruik van Open Spending door journalisten, ambtenaren en politici. Commissioned by Gemeente Amsterdam and Stichting Open State (2014)

29. Author unknown: Wat Doet De Poet? Web application, launched at the Accountability Hack, organised by the Dutch Supreme Audit Institution on 9 June (2017). http://www.watdoetdepoet.nl/. Accessed 20 Oct 2017

30. Dutch Quality Institute Dutch Municipalities: Waar Staat Je Gemeente? Website (2017). https://www.waarstaatjegemeente.nl/. Accessed 20 Oct 2017

31. Alliander. Jaarverslag (2017). https://2016.alliander.nl. Accessed 20 Oct 2017
32. Layne, K., Lee, J.: Developing fully functional E-government: a four stage model. Govern. Inform. Q. **18**(2), 122–136 (2001)
33. Steccolini, I.: Is the annual report an accountability medium? An empirical investigation into Italian Local Governments. Finan. Accountability Manage. **20**(3), 327–350 (2004)
34. Linker, P.-J.: Kordes-Trofee: prijs voor de beste publieke verantwoording. In: Dees, M., et al. (eds.) Externe verslaggeving van publieke organisaties, pp. 229–236. SDU, Den Haag (2009)

Concepts of State and Accounting: A Framework to Interpret the Italian Governmental Accounting Reforms

Luca Bartocci[✉] [iD] and Damiana Lucentini

Department of Economics, University of Perugia,
Via Alessandro Pascoli 20, 06123 Perugia, Italy
luca.bartocci@unipg.it, damianalucentini@tiscali.it

Abstract. The idea of the state, even before having a political connotation, is a concept with a strong philosophical and historical value. According to this assumption, it is possible to conceive government forms as the explicative medium of state power, and the accounting as a kind of technology to put in action this concept. In this sense, is possible to hypothesize a significant link between the evolution of the state models and changes in public accounting. The paper aims to provide a critical interpretation of the Italian governmental accounting reforms in the light of changes of models of state. The case of Italy is very interesting because it is possible to identify some aspects of the traditional model of public administration and, at the meantime, other ones due to the spread of new paradigms for public sector.

Keywords: Accounting history · Governmental accounting · Italy
Paradigms of state · Accounting as a technology

1 Introduction

Over the years the concept of state has undergone a significant change both in relation to political and economic events and their relative impact on society, as well as the effect of its permeation among the different areas of knowledge such as philosophy, administrative science, sociology, political science and economics (Badie and Birnmaum 1984). It is possible to identify the origin of the modern state in the separation between volitional power and enforceable power, considered by some (Carolan 2009) to be also the basis of the modern concept of public accounting.

This paper is based on two affirmations taken as postulates. Firstly, a correspondence between the forms of government and the public accounting system is assumed as existent. Secondly, accounting as considered as a technology, based on the idea that technology is something not exclusively related to the industrial development and even the skills of writing, arithmetic and double entry are a considerable part of technological processes (Parker 1989).

The objective is to create an analytical picture of the parallels between the concepts of state, organizational forms of administration and governmental accounting systems from the perspective of finding the root causes and transformations over time.

© Springer International Publishing AG, part of Springer Nature 2018
T. Antipova and Á. Rocha (Eds.): MosITS 2017, AISC 724, pp. 148–158, 2018.
https://doi.org/10.1007/978-3-319-74980-8_14

Given the specificity of the subject, it is considered important to focus these research hypotheses on a specific case, that is Italy. This choice results from the fact that Italy constitutes an extremely interesting case study for at least three reasons: (a) there is a profound consideration for the evolution of public accounting from a historical point of view, particularly at a central level, (b) the search for an adequate accounting language to be used in public administration is still underway and the government has recently promoted a process of accounting harmonization throughout the entire public sector, (c) significant attention has been reserved over time for governmental accounting – rigorously disciplined by law – which has always been attributed an important role for the workings of public administration.

The paper is structured in other five sections. With the second section a framework to interpret changes in model of the state is provided. In the third section an outline on the evolution of Italian administrative system is presented. Fourth section is about a possible interpretation of the innovations introduced by the different governmental accounting reforms. Finally, some final remarks conclude the work.

2   The Evolution of the Modern State: Old and New Paradigms

The conceptualization of the state and its political and administrative evolution usually follow a classification aimed at highlighting the distinguishing aspects of each era, thus emphasizing the historical links to the social reality. Generally, three State models are singled out and used to illustrate the most significant steps in the transformation of public administration (Borgonovi 1996):

(1) The government of the 1800's which, along with the progressive social ascent of the bourgeois class, sees liberal principles and individual rights being affirmed in an economic sense, logically ratifying the difference between the public and private sector. Ascribing a role essentially as guarantor to such a governmental model, bureaucratic activity ends up being identified with the respect of legality (to be understood as the compliance of the administrative actions with the law).

(2) The concept of the welfare state begins to take form in the 1900's, characterized by progressive legitimization of public intervention in the economic system. The principal causes which can explain this evolution are (Flora and Heidenheimer 1981): (a) socio-economic development, particularly that linked to urbanization and industrialization; (b) political mobilization of the working class, which exercised significant pressure for the acknowledgement of social rights; (c) constitutional development, articulated in two perspectives: the expansion of the right to vote, with the consequent extension of the concept of citizenship, and the virtual appointment of parliamentary responsibility. An additional catalyst for the affirmation of this model is undoubtedly the economic and financial crisis occurred in the second and third decades of the century, which showed the limits of the market in terms of self-regulation and perfect functionality.

(3) The third model, which is still evolving, basically finds its ontogeny in the crisis elements of the welfare model, and primarily in the realization that a merely public management of the services has generated negative economic external effects in terms of inefficiency and ineffectiveness. Therefore, a "hybrid" model is established, defined as a welfare mix characterized by an increase in business-styled management, as well as progressive acknowledgement of organizational and management dictates from the private sector to be extended to the public sectors.

The evolution that follows the welfare mix has been also characterized by the theoretical archetypes of reference, particularly if one looks at the paradigm which has best acknowledged the need for economizing and efficiency as well as a progressive, business-like handling of public administration: the so-called New Public Management (Hood 1989). More recently, up against the increasingly inefficiencies and limitations highlighted by the NPM-inspired reforms, the need for public administration to assume responsibility has arisen as a foundation and criteria for legitimization, not only as far as performance is concerned but also in a more profound and general sense. Thus another ideal type of public administration can be anticipated, which recalls the idea of collectivism in a new and profound sense, in an arrangement that is no longer hierarchical but "relational". From this perspective, in the last years a new perspective has been spreading to correct the distortions caused by the managerialistic wave. In this regards many labels have been used, but one of the most common is New Public Governance (Osborne 2010). As some European authors (Pollitt and Bouckaert 2017) have pointed out, it is also interesting to note the recent return of some elements of the traditional public administration, so that a model known as the Neo Weberian State has been emerging.

3 An Outline of the Evolution of Italian Public Administration

The difficult path to the Italian national unification found its climax in the proclamation of the Kingdom of Italy in 1861. The creation of the national state was a result of the aggregation by the House of Savoy of several pre-existing states, often very small and generally set apart by a secular history. The first legal order of the Italian state was the outcome of the application of the Savoy law inspired by the French experience. Also for this reason, the Italian administrative evolution can be traced back to the tradition of Continental Europe. In this sense, the evolution of the conceptions of state mentioned above can be quite useful to interpret Italian institutional history, taking into account some distinctive elements.

The cultural boost propelling towards the creation of a new unified state is, for the most part, attributed to the enlightened and Jacobean bourgeoisie. The rapid pace of the Italian unification process (just a few decades) leaves the country with serious problems of uniformity and integration. Due to centuries of political fragmentation the Peninsula find itself with enormous ethnic, geographic, historic and socio-economic differences.

A long period of institutional adjustment characterizes the first decades of the new state. The following features can be identified:

- the government's desire to assume a central role to the detriment of parliamentary power, while the traditional state of rights imply central legislative body and a more strictly executive role of the governments;
- in the restructuring process of the administrative machine two important criteria are adopted: uniformity and centralization;
- the effective lack of rival powers superior or inferior to the State, and also the influence of organizations created in the civil society was very limited;
- the emphasis of preventive administrative and accounting controls, laying the foundation for bureaucratic formalism which the Italian public administration still suffers from;
- the ever growing dependence of public finance on the tax system caused by the gradual alienation of large real estate and land properties acquired by the state during the unification process;
- along with the typical functions of the so-called minimal liberal state (foreign affairs, army, police and legal authority), attention is also placed on certain areas of a strong social value: infrastructure and public education.

This constituent phase of the institutional and administrative machine inevitably includes public accounting rules. The first regulatory reference on the subject is the introduction of the 1869 reform. This provision, established by the General Accounting Office (GAO) and the central accounting offices within each ministry, dictates the rules for estate management and the activities of the accounting system. The main documents include a budget statement and a financial report which the government annually presents for parliament approval.

At this point it is important to point out:

- the division of income and expenditures administrations (Finance vs. Treasury Ministries);
- the adoption of cash and obligation basis accounting (a sort of modified cash accounting) which provides both cash and obligation based budget;
- the use of budget as a tool for authorization and limitation of expenditures;
- providing the general financial report with an identical financial statement and an assets and liabilities statement.

This provision is further changed with a new regulation of 1883 which permanently defines the centrality of obligation based budget.

By the end of the 19th century and the beginning of the 20th, once the adjustment period has ended, it is again possible to find most typical characteristics of the liberal state in Italy. Later, when the Fascist government came to power after the World War I, a strong consolidation and successive expansion of the public machine occurred.

The will of an increasingly more tightening control of politics on administrative action is an interesting interpretation for the subsequent accounting reform passed into law in 1923. In brief, it has the following provisions:

- ministerial accounting offices are placed under the GAO with a centralization of power for accounting control (especially preventive control);
- the budget authorization and limitation function is intensified;
- more importance is given to obligation basis accounting, accompanied by a cash budget, which is updated on a quarterly basis for constant control of expenses;
- an attempt is made to estimate a final balance taking into account monthly summary accounts for annual reporting.

In a neutral public finance scenario, where bureaucracy is perceived as a operational tool of the political power, the idea of accounting as a means of guidance and control of administrative action is reinforced, with a special attention to expenses.

The '30s signal a turning point in the formulation of economic policy in the Fascist government. Different factors, including the sensational failure of the markets, certainly push different countries to reconsider the economic role of their own public institutions. In Italy all this is embedded in an ideological concept that renders the nation-state identity one of its cornerstones. Economic autarchy becomes a key-value in the government policy. This tendency lasts until the World War II, after when Italy becomes a republic. The State's institutional set-up changes and a new system with a more rigorous division of powers emerges with a more substantial affirmation of parliament centrality.

Also the idea of public finance changed, and government spending becomes the means of wealth distribution. Furthermore, the way financial statements are drawn up changes. A new law passed in 1964 conceives the budget as an act of political guidance and a tool of economic policy. It modifies the budget structure and introduces criteria for item classification to show the impact of public finance on the socio-economic system. This prospect is then consolidated by the subsequent 1978 reform which acknowledges the logic stemming from the American PPBS (Planning, Programming and Budgeting System) approach. The substantial difference in this provision is a series of tools aimed to set up an articulate process of economic and financial planning. In essence, this includes:

- a medium-term, political guidance document;
- a long-term budget explaining the actions in terms of three-year previsions;
- a descriptive narrative document including the entire planning process.

The 1978 reform, with the modifications in 1988, represent the peak of a functional concept of public finance and that of the budget. The budget parliamentary debate became the most delicate moment in the political and economic life of the country. From an administrative point of view, the process of fragmentation began.

In the '90s, Italian public administration enters a new renewal phase which continues up to the present time. This period can be defined as an ongoing reform. The changes affect all levels of public administration and aim at renewing practically all its aspects. It is interesting to note that the initial stage of this reform process is generally referred to as *aziendalizzazione* of public administration (literally transforming into a business). It is characterized by more result-oriented principles and typical instruments of private management. The main driving forces behind this change were the reforms of internal controls allowing strategic and managerial control, as well as the renewal of accounting systems, aiming to find more meaningful accounting information for analysis and evaluation.

In 1997, a new governmental accounting reform came into force. Its main ideas can be summed up as follows:

- the creation of two distinct budget levels: a political one, to be submitted to a parliamentary vote, and a managerial one, entrusted to the top management;
- an attempt to streamline parliament decisions and make them more conscious, singling out a new accounting voting unit ("basic budget unit");
- the introduction of zero-base budgeting method;
- assigning responsibility and autonomy to directors to manage their own budgets;
- the creation of assumptions to implement tools for performance evaluation;
- making the budget more understandable to the citizens.

The main assumptions of the reform are the distinction between the guidance role of the political organs and the technical tasks of bureaucratic organs, and the adoption of management-by-objectives model.

This great renewal will have to produce the desired effects, however it is certain that it will require an additional, second phase of the reform. Not by chance, in 2008 the GAO started to experiment with a new budget to align the budget structure closer to the United Nation COFOG (Classification of the Functions of Government) model. The new classification of items is focused on missions and programs. Subsequently, in 2009 a new law approving fiscal federalism gave the government the right to legislate the harmonization of accounting procedures for all public sector. In the meantime, by the end of 2009, the parliament approved another bill regarding governmental accounting system. This provision makes the GAO practice official and renews considerably the financial planning process. Indeed, the Italian public administration can be still compared to a huge open construction site.

4 Interpreting Governmental Accounting Reforms

After having gone through the major stops along the path of Italian public administration, an interpretive plan of the relational factors which have conditioned the reformist choices regarding governmental accounting is provided.

On the basis of the main passages reported, it is possible to superimpose the suggested timelines in order to obtain a representative plan, obviously simplified, of the evolutionary process in question (Fig. 1).

This outline allows three important historical moments to be identified.

The first one, which we can conventionally place from the time of national unification through to the 1930s, corresponds with the period of cultural and political affirmation of the state as an Institution, a unifying subject breaking away from the state as a community. In fact, the executive body taking a protagonist role in addition to the centralization of administrative activity can be seen here. As the structure gradually

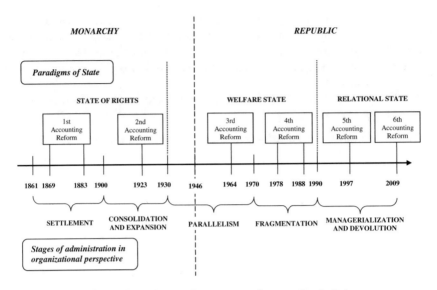

Fig. 1. The reforms of governmental accounting in Italy

settles down, the result is an ever-increasing expansion of the bureaucratic machine. The fundamental mission of political power is to pursue substantial unification through the logic of standardisation. A strictly internal perspective is assigned to the accounting system, the subject of two substantial reforms, essentially performing a regulatory function in the relationships between parliament and the government and between the institutions and the bureaucratic structure themselves, as a sort of bind. The function of administrative control of the budget (especially expenditure), becomes more marked with the Fascist Reform of 1923.

The second period covers the '30s to the '80s and is characterized by the emergence of the Italian version of the Welfare State in which public power internalizes the primary requirements of the less well-to-do and becomes an immanent subject of the socio-economic system, no longer an arbiter but a "providence" for society. These decades see the transition from the monarchy to a republic and, in a quest for the most protective institutional structures, parliamentary centrality is re-established. State interventionism is made possible through forms and entities competing with the more traditional administrative action. From the '70s onwards, this "administrative paral-lelism" would result in an excessively cumbersome and complex structure which would begin to disintegrate without an organized plan. In this long phase, the budget acquires macroeconomic significance becoming the principal tool of economic policy. With the 1978-1988 reform, policy decision should be the logical outcome of an articulated debate between the government and parliament. The resulting effect is an encumbering of the path leading to the budget which is becoming more and more burdened with extra-financial aspects (macroeconomic interventions, industry restructuring, specific measures for territories or subject types). The budget becomes an instrument of political and social consensus, above all with reference to the equilibrium between the parties and relationships between the government and social forces (consider the importance assumed by the unions during these years).

The third period which can be said to begin in the '90s, brings us to present-day discussions. The State is going through a phase of reflection regarding its role and its "way of being". Initially, organizational and management renewal is pursued using certain private sector managerial models as a point of reference. The public sector, whose burden is eased by the initiation of the privatization process of certain companies and management of specific services, begins to be seen as a network made up of several administrative units. The perspective becomes microeconomic and attention is directed particularly towards the production techniques/provision of services. Recently, a broadening of scope has been promoted through two directives: a shift in focus from the issues of efficiency and effectiveness to the broader one of public value (Moore 1995) and the interaction of more subjects (Stoker 2005), both public and private. From an accounting point of view, the break from the past comes as a result of the 1997 reform: with the new budget an attempt is made to streamline parliamentary decisions while managerial budgets for individual Ministries are required to be drawn up and managed according to business logic and techniques. For cognitive purposes, furthermore, the presentation of the accrual budget in addition to the traditional budget is introduced. The situation is still evolving: important innovations are foreseen stemming from the latest reform, n. 196, dated 31 December, 2009. The initial intent appears to be that of streamlining the planning process, thus regaining its financial nature. At a moment in which important shares of revenue and buying power are about to be decentralized, budgets are set up to carry out a regulatory function in public finance. From this point of view, in other words, the budget would take on a state role as coordinator and agent against the possible tendencies towards break-up given the obligations resulting from joining the euro and the unifying boost received from the application of fiscal federalism and an autonomous finance model. In light of this interpretive layout, it is now possible to concentrate more specifically on the various reforms, bringing together the functions which can be predominantly recognized by an accounting system and the technical options used.

Considering the main categories of subjects expecting to receive information, we can draw up the following table (Fig. 2).

The outline shown above illustrates how the meaning predominantly assigned to the budget has undergone numerous changes over the years, gradually assuming more political, legal, macroeconomic, managerial, financial connotations, depending on the strength ratio between the various interest centres. It is also true that such values are not mutually exclusive and that a stratification of meanings may be recognised. This ever-increasing multi-functionality does not correspond to a similar evolution in the accounting basis used.

5 Final Remarks

The analysis undertaken confirms the two initial assumptions: (a) a correlation exists between the state paradigm, prevalent during a specific historical period and the characteristics of the state accounting system, particularly with regard to the structure

Accounting Reforms	Primary significance of budgetary process	Information logics of budget structure	Accounting basis for budgeting	Financial planning tools
1st Reform: - L. 5026/1869 - R.D. 5852/1870 - L. 1455/1883	Political bind and financial checks and balance	Ordinaries and recurrence of expenses	(Cash till 1883) Cash and Obligation	—
2nd Reform: - L. 1601/1922 - R.D. 2440/1923 - R.D. 827/1924	Administrative limits	Ordinaries and recurrence of expenses	Cash and Obligation	—
3rd Reform: - L. 62/1964	Economic policy tool	Economic nature and function of expenses	Cash and Obligation	—
4th Reform: - L. 468/1978 - L. 362/1988	Political synthesis	Economic nature and function of expenses	Cash and Obligation	• Economic & Financial Planning Document • Forecasting Planning Report • Finance Law • Multi-year Budget
5th Reform: - L. 94/1997 - D.Lgs. 279/1997 - L. 208/1999	Negotiation with management	Administrative responsibility	Cash and Obligation	• Economic & Financial Planning Document • Forecasting Planning Report • Finance Law • Multi-year Budget
6th Reform: - L. 196/2009 - L. 39/2011	Regulation of decentralized finance in compliance with EU rules	Missions and programs	Cash and Obligation	• Report on the Economy and Public Finance • Decisions Regarding Public Finance • Stability Programme Update • Stability Law • Multi-year Budget

Fig. 2. The reforms of governmental accounting: an analytical outline

of the budget; (b) accounting is a technology able to regulate mutual relationships between subjects inside and outside public administration.

It should be noted that in the Italian experience, legal planning of the accounting procedure has been judged the best way to implement uniformity and a change in organizational behaviour. In this sense, accounting reforms are not only the expression of a certain way of understanding the government of "public affairs" but they are also the primary means with which such political-cultural paradigms are spread.

There is no doubt, therefore, that public accounting (and, in particular, the budget) are important political vehicles and that the government has always played a leading role in the system of relationships considered.

Faced with a substantial temporal standardisation of the accounting basis, the aspect which has proved to be most sensitive to changes in the "information expectations" is the formal structure of the annual budget. In particular, what has changed especially over time is the logic of classification of financial size and the elementary accounting unit submitted to the parliament for voting the budget.

Briefly it may be affirmed that:

– when the relationship of the main force is modified, following a change in the politico-administrative scenario, the reason why the budgeting process was conceived in fact changed and this led to subsequent reviews of the accounting system;

- such changes (cultural before becoming regulatory) have not led to significant variations in the accounting basis since the latest reform has opened up the way to an unexpected return to cash accounting. Immediately after the initial period cash and obligation basis accounting has always been considered the best information base when programming and controlling administrative activity since it takes into consideration not only cash flow but also any commitments undertaken and rights matured in their favour;
- faced with this assumption, the "information pressures" go in two directions: regarding the formal structure of the documents, an urgent request for a search for more expressive forms of statement; secondly, there emerged a tendency to release information documents which are added to those concerned in the parliamentary vote.

After years of debate regarding the prospect of adopting forms of accrual basis accounting also in the public sphere, the Italian legislator has embarked on a path which seems to be imposed by the need to "keep a tight rein" on peripheral administrations. There is no doubt that, though it is comprehensible, the choice made could be controversial and could lead to questions about the degree of effective assimilation of some basic principles of the NPM (Lapsley 1999) by the Italian political and technical world (Mussari 1997).

The historical path proposed illustrates the particular difficulty encountered by public accounting in freeing itself from bureaucratic formalism. Accounting is not only an "obligation" or a "technical matter". Despite the fact that a large number of Italian politicians and bureaucrats continue to see it only in this way, it is a tool of knowledge, of awareness, of responsible information exchange, of sharing.

References

Badie, B., Birnmaum, P.: The Sociology of the State. Chicago University Press, Chicago (1984)

Borgonovi, E.: Principi e sistemi aziendali per le amministrazioni pubbliche. Egea, Milano (1996)

Carolan, E.: The New Separation of Powers: A Theory for the Modern State. Oxford University Press, Oxford (2009)

Flora, P., Heidenheimer, A.J.: The Development of Welfare States in Europe and America. Transaction, Piscataway (1981)

Hood, C.: Public administration and public policy: intellectual challenges for the 1990s. Aust. J. Public Adm. **48**(4), 346–358 (1989)

Lapsley, I.: Accounting and the new public management: instruments of substantive efficiency or a rationalising modernity? Financ. Account. Manag. **15**(3–4), 201–207 (1999)

Moore, M.: Creating Public Value. Harvard University Press, Cambridge (1995)

Mussari, R.: Autonomy, responsibility and new public management in Italy. In: Jones, L.R., Schedler, K., Wade, S.W. (eds.) Advances in International Comparative Management - International Perspectives on the New Public Management. Jai Press Inc., Greenwich, Connecticut (1997)

Parker, R.H.: Accounting basics: language, writing materials, numerals and calculations. In: Macdonals, G., Rutherfort, B.A. (eds.) Accounts, Accounting and Accountability. Van Nostrand Reinhold, London (1989)

Pollitt, C., Bouckaert, G.: Public Management Reform: A Comparative Analysis into the Age of Austerity. Oxford University Press, Oxford (2017)

Osborne, S.P.: The New Public Governance? Emerging Perspective on the Theory and Practice of Public Governance. Routledge, Oxon-New York (2010)

Stoker, G.: Public value management: a new narrative for networked governance? Am. Rev. Public Adm. **36**(1), 41–57 (2005)

Governmental Auditing Systems
in Indonesia and Russia

Tatiana Antipova[✉]

Federal State Budgetary Educational Institution of Higher Education
"Perm State Agro-Technological University Named After Academician
D.N. Pryanishnikov", Institute of Certified Specialists, Perm, Russia
antipovatatianav@gmail.com

Abstract. This paper compares Indonesian and Russian governmental auditing systems. The data of this study comes from Governmental auditing Institutions' reports. The research result lightens the most widespread similarities and differences in Indonesian and Russian governmental auditing systems. The data of this study comes from Governmental auditing Institutions' reports. The research result lightens the most widespread similarities and differences in Indonesian and Russian digital governmental auditing systems. When comparing governmental auditing systems, their structure, goals and objectives are shown.

Keywords: Governmental auditing system · Accounts Chamber
Supreme Audit Institutions · Control procedure
Objects' governmental auditing · Subjects' governmental auditing

1 Introduction

Nowadays, with globalization of the economy, good governance of public finance is of fundamental importance for all countries to ensure the sustainability of the national budget and financial systems, as well as mutual financial security and sustainable economic growth. The public sector of many economies has been subject to dramatic change. Transparency and public accountability further engender trust in a representative democracy. Working together, these factors lead to greater citizen satisfaction and better access to capital at a lower cost. As long as budget resources come from the public in the form of taxes, every citizen demands greater understanding of where their tax money goes and how it is spent and control.

Governmental auditing focuses attention on how budget resources are spent. Assessment procedures of governmental auditing serve to avoid misrepresentation and fraud in public sector financial statements [1]. Since the public sector financial statements are placed on the Internet, any citizen can get acquainted with this reporting. And citizens can trust the information set in these financial statements if it is verified by auditors properly. Governmental Auditors should assess fair presentation of public sector financial statements. When financial statements "furnished by a state agency are in fact reliable, citizens' trust should be increased by auditors. When, on the other hand the information is significantly unreliable, auditors should reveal that and consequently decrease citizens' trust" [2].

© Springer International Publishing AG, part of Springer Nature 2018
T. Antipova and Á. Rocha (Eds.): MosITS 2017, AISC 724, pp. 159–166, 2018.
https://doi.org/10.1007/978-3-319-74980-8_15

To understand the essence of governmental audit, it is necessary to define its subject and object. The subject is the audit institution and the object is the auditee. The subject depends on the public sector level.

At the central government level, financial statements are audited by federal government or parliamentary bodies like Supreme Audit Institutions (SAI), federal financial control bodies, treasury, Accounts Chambers, etc.

At the state/regional government level, financial statements auditing (control) is conducted by regional branches of federal financial control bodies, state Accounts Chambers, regional Account Chamber.

At the local government level, financial statements auditing (control) is conducted by local parliamentary bodies like local Accounts Chambers and some of central and state governments' body in sharing functions.

The object is auditee – mostly public sector entities belonging to the public sector that present financial statements.

Financial statements of an entire country presented to the International Monetary Fund (IMF) are prepared and approved by the member country's Ministry of Finance in accordance with Government Financial Statistics Manual (GFSM, 2014). All 188 IMF member countries (including Indonesia and Russia) must comply with GFSM 2014. To conduct governmental auditing in the public sector, auditors must know public sector accounting standards, rules, laws, and regulations for preparing financial statements, as well as auditing standards, rules, laws, and regulations for writing auditor's reports. Because "the auditor's report is the most important product of the audit. The auditor's report is normally directed to parliament and the cabinet. It is thereby often made available to citizens." [2].

The report should include all significant instances of noncompliance and abuse and all indications or instances of illegal acts that could result in criminal prosecution that were found during or in connection with the audit (GAO/OP-4.1.2).

When giving the auditor's report, auditors must be sure that financial statements are complete, reliable, accurate, consistent, and timely. In addition to law, "the rules and regulations specifically relevant to audit in the public sector can be presented in four categories: the Lima declaration, the Code of Ethics, auditing standards and practice notes" [2].

First of all, it is necessary to know the structure of the public sector. The most common public sector comprises general government and public corporations. Government-owned enterprises, such as the central bank, post office, or railroad are often referred to public corporations. General government usually consists of three levels: central government, state or regional government, and local government. Public corporations are divided on nonfinancial (e.g., post office) and financial (e.g., central bank).

2 Indonesian Governmental Auditing Outline

Indonesia is one such developing nation; it has undergone dramatic social, political and economic reforms. Within Indonesia, given its long colonial history, a dialogue among stakeholders is not a tradition to which its long-enduring public is accustomed.

Furthermore, Indonesia's adoption of the new public financial management may not nurture the sort of public engagement it experiences in nations with democratic traditions. Whether and how the Indonesian reform has opened up opportunities for local dialogue is, thus, worthy of study [4]. At the same time on the official web site (http://www.bpk.go.id/) name of BPK RI is the Audit Board of the Republic of Indonesia. But according to Indonesian Constitution of the Republic of Indonesia, 1945 (Art. 23E), this name must be Supreme Audit Board. To avoid misunderstanding there is using Indonesian abbreviation - BPK RI.

BPK RI has the authority to examine the reports of all government agencies and state-owned firms. Audit reports that indicate unlawful practices must be reported to police or attorneys. Audit reports are now available to the public. The authority of BPK RI was also strengthened, giving it the right to audit government reports, public sector spending, and any government projects. The only parties with access to BPK RI opinions in the past were president-approved members of National parliament; audit opinions were treated as state secrets [4]. BPK RI became independently empowered and was required to report to the broad public (Law 15, 2006, Article 2). Audit opinions and (now) recommendations from the Board were to be used by Parliament to evaluate performance at all levels of government (Central, Provincial, and Municipal). Notifications from BPK RI indicating fraud or misuse of resources (common in Indonesian government practice, [4], were to be reported to the police or attorneys general, and were applicable to all levels of government (Law 15/2006, Article 8).

All central and Local Governments are now required to offer accrual-based balance sheets, income and expense reports, and performance reports, in addition to the traditional cash flow and budget statements (Law 17/2003, Article 33). Local authorities must employ a performance based budgeting system (Law 25, 2004, Articles 4 and 5), which takes into account Central Government's strategic plan (Law 15/2004, Sects. 4 and 5). A national executive body – including elected officials – must now consult with Parliament and the public to set programmes and budgets. Such consultations must be made available under formal public meetings (Law 24/2004), and costs must be subject to local and public review. The potential for local voice, concerning fiscal matters of government, is now legislatively empowered, and opportunities for public discourse are required under law. Thus "voice" has been facilitated for these local players, at least in theory. Anyone can now access financial and audit reports as well as criminal and fraud report information from BPK RI. [4].

Auditors are required to follow the information technology (IT) development when performing the audit work [3]. Computer Assisted Audit Techniques (CAATs) is divided into three (3) approaches: the computer audit, the audit through the computer, and computer assisted audit. The concept of e-audit in the public sector has been included in the strategic plan of BPK RI 2011–2015 in the framework of inspection management and financial responsibility of the state [6]. This concept is planned to address the problems associated with the short inspection time and the limited number of auditors due to the increasing number of state entities. Application of e-audit concept in the public sector is expected to provide benefit in preventing, detecting and tracking the fraud in the management and accountability of state financial (BPK RI, 2012). BPK can employ external examiner such as a public accountant who works in the firm who is registered in BPK to assist the state financial examination duties (Act No. 15 of 2004 Chap. 9).

Auditor's performance increased due to the implementation of data collection and inspection process by using e-audit was found to be faster than the conventional way. Advances in technology could make the auditors to ensure the internal control of the auditee, to access documents and records as well as produce a more efficient information that can not be performed by using manual audit approach. If e-audit implementation has an impact on improving the auditor's performance, then increased performance will automatically raise the audit quality and the reports it generates [6].

BPK RI can compile and submit the Semester Examination Result Overview (IHPS) each fiscal year to the representative agencies and government on time. IHPS is a document that contains a summary of BPK RI audit results, noncompliance with regulatory requirements Legislation, follow-up monitoring recommendations, result of examination, and result of settlement monitoring of state/region loss for half a year. IHPS is structured to fulfill the mandate of Law Number 15 Year 2004 regarding Audit of Financial Management and Responsibility State, Article 18. Under the aforementioned provision, the Supreme Audit Board shall be required to submit the IHPS to Representative institutions and the President/governor/regent/mayor at the latest three months after the end of the term. BPK RI identified a number of reasons for non-compliance with regulatory requirements Legislation, including a lack of coordination within the government at all levels, a lack of skilled employees, and rapid changes in regulations.

For the purposes of current study we analyzed BPK RI audit results comes from IHPS for three years: 2013–2015. In more detail, the I Semester Year 2015 results of the examination on local government and BUMD revealed 8,019 findings containing 12,170 problems. These problems include 6,034 (49.58%) weaknesses of SPI and 6,136 (50.42%) noncompliance with the statutory provisions of Rp14.39 trillion. From the problem of non-compliance, there are problems of financial impact as much as 3711 (60.56%) worth Rp11, 90 trillion.

IHPS II Year 2015 compiled from 6,548 findings containing 8,733 Problems, covering 2,175 (25%) weaknesses of SPI and 6,558 (75%) The issue of noncompliance with the provisions of legislation Worth Rp11.49 trillion.

These financial impact 2014 issues consist of non-compliance to the provisions of legislation resulting in losses of Rp1.42 trillion, potential losses of Rp3.77 trillion, and a shortfall in revenues of Rp9.55 trillion. In addition, there are 3,150 problems of ineffectiveness, inefficiency, and ineffectiveness worth Rp25.81 trillion.

In the second semester of 2013 BPK has disclosed 10,996 cases worth Rp13.96 trillion. Of these, 3,452 cases worth Rp9.24 trillion are findings of non-compliance resulting in losses, potential losses, and lack of revenues. During the review process, BPK has saved the money/state assets derived from the transfer of assets or deposits of money to the state/region/state-owned/region-owned enterprises valued at Rp173.55 billion derived from the entity's follow-up to non-compliance findings (Ikhtisar Hasil Pemeriksaan Semester II Tahum 2013. BPK RI, Jakarta, Maret, 2014).

3 Russian Governmental Auditing Outline

In Russia, the analogue of the government audit is the state financial control. Russian state financial control is designed to detect only those cases of infraction, abuse, or program incompliance that can be evidenced by primary documents. Thus, preferably there is carried out follow-up control that the non-compliance with regulatory requirements legislation state ex post. In Russia, the terminology of that control process is not approved by federal law and in different agencies' instructions are referred to differently. But in the Budget Code, the basic law regulating the budget process in Russia, the term 'state financial control' is used, although there is not an approved definition for this concept (see Federal Law "Budget Code of Russian Federation" from 31.07.1998 №145-FZ). This term is does not defines neither who in the government or banks has the power to control budgeting issues, what the order of control is. There are no common governmental auditing standards for Russia as a whole. All control procedures of state financial control in Russia are performed by inspectors, not auditors. Russian governmental auditing systems consist of external auditors mostly.

In general Russian governmental auditing systems consist of three levels: federal, regional and local. On federal level the most significant Governmental auditing Institution is Accounts Chamber of the Russian Federation, at the regional level there are 85 regional Accounts Chambers, and on the local level there are just some municipal Accounts Chambers. All of Russian Accounts Chambers belong legislative power branch and independent from executive branch theoretically. But inherently its funded by the appropriate level Ministry of Finance that belong to executive branch of government power.

The main state bodies exercising control of the federal budget, are the Accounts Chamber of the Russian Federation (hereinafter – Accounts Chamber RF). The Chair of Accounts Chamber RF is nominated by the President, and then approved by Federal Assembly. In addition, the Accounts Chamber reports to the Federal Assembly. Accounts Chamber operates under the authority of the legislature.

According to Federal law from 05.04.2013 N 41-FZ "Accounts Chamber of the Russian Federation", Accounts Chamber RF performs many functions.

One of the most important tasks performed by Accounts Chamber RF is to analyze shortcomings and violations in the process of disposal of federal resources. Therefore, it is of interest, how to relate to the total amount of violations and the total amount of federal spending. The most serious financial violation in Russia is the no-purpose use (misuse) of budgetary funds, for which may be applied criminal liability. Other types of violations (mostly non-compliance with regulatory requirements legislation) are classified year to year differently depending on control body internal policy.

For the purposes of current study we analyzed Accounts Chambers RF governmental audit results comes from Annual Statements for three years: 2013–2015. According to Accounts Chambers RF Annual Statements in 2013 were revealed 22087 million Rubles of financial violations (non-compliance with regulatory requirements legislation); 2014–9323 mln Rub; 2015–7086 mln Rub respectively.

In following paragraph we compare Indonesian and Russian governmental auditing statements' results for 2013–2015.

4   Indonesian and Russian Governmental Auditing Systems Comparison

The most important and common thing is that Indonesia and Russia are members of the IMF. Both countries must to comply with IMF rules related to public sector finance including governmental auditing. Also both countries have the same period for counting a fiscal year lasts from 01 January to 31 December and equal to calendar year. Comparison of government auditing system in Indonesia and Russia shows in Table 1.

Table 1. Comparison structure of Indonesian and Russian government auditing systems' levels.

Indonesia		Russia	
Level of authority	Appropriate governmental auditing bodies	Level of authority	Appropriate governmental auditing bodies
National	Supreme audit board (BPK RI)	Federal	Accounts Chamber of the Russian Federation
Local government (Provincial)	34 Regional branch of BPK RI, Provincial Supreme audit boards	Regional	85 regional Accounts Chambers
Local (municipality)	Local Supreme audit boards	Local (municipality)	Municipal Accounts Chambers

To compare the essence of governmental auditing in Indonesia and Russia was considered activity of main control body in each country: BPK IR and Accounts Chamber RF. Their objectives compared in Table 2.

Table 2. BPK RI and Accounts Chamber of RF objectives' comparison.

BPK RI objectives	Accounts Chamber of RF objectives
Central Government/Local Government, Institutions of Other Countries, Bank Indonesia, State-Owned Enterprises, Agency Public Service, Regional Owned Enterprise, Foundation, and Institution or Other bodies	Public Sector Institutions, Bank of Russia, Non-Profit Organizations, Commercial organizations, Individuals, Any bodies, organisations, and individuals under investigation by law enforcement authorities

During current study author analysed official statements of BPK IR and Accounts Chamber RF set on their official web sites in open access. For the purpose of comparability, annual reports in national currency are converted to data in million dollars. The dollar exchange rate was taken on December 31, end of each fiscal year. Currency comes from http://www.bi.go.id/en/moneter/informasikurs/referensi-jisdor/Default. aspx and http://www.cbr.ru sources respectively. For example, the 2015 reporting data are converted to million USD from Indonesian Rupee (IRD) and Russian ruble

(RUB) rates on December 31, 2015. Analysis result for three fiscal years converted governmental audit reports' data present in Table 3. Data come from IHPS of BPK RI and Annual Statements of Accounts Chamber RF.

Table 3. Comparison of the sum of detected non-compliance with regulatory requirements legislation, revealed for the fiscal year by BPK RI and Accounts Chamber RF, million USD.

Fiscal year	The sum of detected non-compliance with regulatory requirements legislation, revealed for the fiscal year	
	BPK RI	Accounts Chamber RF
2013	1 957	22 087
2014	2 075	9 323
2015	1 876	7 086

After conversion into million dollars we can see that the magnitude of the non-compliance sum revealed for fiscal year is comparable. But the unpredictable fluctuations of these values indicate the weakness of methodical maintenance of control procedures in both countries.

5 Conclusions

Now seems like an opportune time to improve governmental auditing in Indonesia and Russia that is dealing with some major headwinds across a variety of fronts. Many types of crisis, government failures, bureaucratizing, and budget troubles called for a new innovative approach to Indonesian and Russian governmental auditing. As a result of this study, we reached to the conclusion that there are some similarities in the Indonesian and Russian governmental auditing systems: three level systems, the same fiscal year period, similar goals and objectives, comparable audit results. Both countries are member of IMF but nor Russian either Indonesian governmental auditors have implemented GFSM 2014 rules into daily practice. In addition, it is necessary to note that "many auditors want further hands-on guidance to be included in the standards" [2]. Despite the fact that Indonesian and Russian auditors have many types of instructions, there are still not enough really workable, clear, and easy manuals for beginners and students. These manuals may lead to increasing methodical maintenance of control procedures in both countries. Both Indonesian and Russian governmental auditing needs to rethink its priorities, put transparent decision-making, simpler legislation, make better methodology, and simplify the rules it uses to define and measure auditing results.

To improve auditing procedures Indonesian and Russian governmental auditors should use modern information digital technology; strengthen investigative powers; encourage more professional designation; support the international transparency. Moving auditors to new and evolving techniques will modernize financial statement auditing by making full use of current and emerging technologies to overhaul traditional sampling-based auditing approaches and fully leverage technology to digital auditing. Digital auditing allows leveraging sophisticated tools, such as online analytical processing to analyze large populations of both manual and automated journal entries from the financial management system. Getting there will take dedicated investments, concerted effort, executive-level commitment, and strong partnerships with agency management, who will likewise greatly benefit from this evolution. In doing so, it will enable the auditor's profession to move into the future and add even greater value to managing the cost of government and providing the highest levels of accountability and transparency to the public [5].

References

1. Antipova, T.: Auditing for financial reporting. In: Farazmand, A. (ed.) Global Encyclopedia of Public Administration Public Policy and Governance. Springer, Cham (2016). https://doi.org/10.1007/978-3-319-31816-5_2304-1
2. Budding, T., Grossi, G.: Public Sector Accounting. Taylor & Francis, New York/London (2014)
3. Fleenor, W.C.: Implication on computers in financial statement audits. J. Account. **179**(4), 91–93 (1995)
4. Harun, H., Van-Peursem, K., Eggleton, I.R.C.: Indonesian public sector accounting reforms: dialogic aspirations a step too far? Account. Audit. Account. J. **28**(5), 706–738 (2015)
5. Lewis, A.C., Neiberline, C., Steinhoff, J.C.: Digital auditing: modernizing the government financial statement audit approach. J. Gov. Financ. Manag. **63**(1), 32–37 (2014)
6. Maria, E., Ariyani, Y.: E-commerce impact: the impact of e-audit implementation on the auditor's performance (empirical study of the public accountant firms in Semarang, Indonesia). Indian J. Commer. Manag. Stud. **5**(3), 1–7 (2014)

Information Technology in Business and Finance

The Virtual Reconstruction of the Earliest Double-Entry Accounting Ledger

Mikhail Kuter$^{(\boxtimes)}$ (iD), Marina Gurskaya (iD), Angelina Andreenkova (iD),
and Artem Musaelyan (iD)

Kuban State Univesity, 149 Stavropolskaya Street, 350040 Krasnodar, Russia
prof.kuter@mail.ru

Abstract. The present paper contributes to the history of the accounting development as this research has allowed us to recreate the initial numeration of the pages in the General ledger of Giovanni Farolfi's company in Provence (1299–1300) using the methods of logical-analytical reconstruction and cross-references as well as to accomplish the virtual reconstruction of its structure. The account book of interest can be truly concerned to be the earliest accounting ledger kept according to the rules of double-entry bookkeeping. The implemented study and its results do not often coincide with those ones described in the works of the previous researches.

Keywords: Branch office of the company of Giovanni Farolfi in Salon
Double-entry bookkeeping · Cross-references
Logical-analytical reconstruction · Virtual reconstruction

1 Introduction

In recent seven decades one of the earliest preserved ledgers of double-entry book-keeping (probably even the earliest one) is considered to be the Ledger of the branch office in Salon (1299–1300), "a town in the independent county of Provence, about 45 miles from Nimes and 25 east of Arles, with a population in 1300 of some 2,000 to 3,000" (Lee 1977, p. 80). The branch office pertained to the company of the Italian merchants in Nimes.

As it is known out of all the accounting registers that are included into the accounting complexes (related to the previous, current and the subsequent accounting periods), this is the only account book preserved. On top of that, the major part of the pages of this Ledger was lost and also the sequence of the available pages was broken. The situation is overburdened by the fact that the upper right corners of the pages are torn or virtually unreadable. They hold much significance as the medieval accountant (whose name was Amatino (or Matino) Manucci, one of the partners who acted as a bookkeeper) numbered the folios there.

The paper aims to identify each page preserved, to reconstruct the historical page numeration and to reproduce the sequence of the preserved pages of the ledger, that was set by Matino Manucci more than seven centuries ago, by means of its virtual computer reconstruction. In the furtherance of this goal we have analysed the cross-references of each entry and applied the method of local-analytical reconstruction (MLAR).

© Springer International Publishing AG, part of Springer Nature 2018
T. Antipova and Á. Rocha (Eds.): MosITS 2017, AISC 724, pp. 169–184, 2018.
https://doi.org/10.1007/978-3-319-74980-8_16

MLAR is intended to bind all the transactions, registered in all the journals (accounting books) that are included into the archival accounting complex, into one diagram. Its basis is the principle of decomposition of the indicators (vertically down), reflected in the Trial balance (Balance sheet) and in the "Profits and Losses" account. There used the principle of composition (aggregation) of the indicators (vertically up) in order to study the early stages of accounting (before the use of Trial balance (Balance sheet) and "Profits and Losses" account).

2   Review of Prior Literature

Before proceeding to the analysis of the references of the authoritative researchers, who wrote about the G. Farolfi's branch office in Salon, it is necessary to provide the paper with the convincing arguments to prove that the studied archival artefact is indeed the earliest preserved monument of the double-entry bookkeeping application. This should not be confused with the earliest saved experience of using a double entry (the fragments of the book of the Florentine bankers at the fair of San Broccolo near Bologna, 1211).

Fabio Besta, who was the founder of the school of science associated with the accounting history (Besta 1909, 1922), as well as his disciples (Alfieri (1911), Rigobon (1891a, 1891b)) considered the books of the Venetian trader A. Barbarigo to be initially the earliest ones, as they completely fell in line with the list of formal features of double-entry bookkeeping, specified during numerous discussions of the first half of the 20th century: unified money measurement; dual accounts, divided into sections of debit and credit; a double entry containing cross-references to the corresponding account; the existence of special accounts for the accounting of revenues and expenses and intended for the calculation of the operational results; the presence of the "Profits and Losses" account for the formation of the financial result on all kinds of activity; the application of a balance sheet to control the settlements; informing the owners and other concerned parties (within their competence) of the financial situation and the financial result.

A little later, the priority was given to the book of the massari (financiers) of the Commune of Genoa (1340). The authors have studied this archival artefact in detail using MLAR (Kuter et al. 2013).

The only scholar of Besta, – Ceccherelli (1913, 1914a, 1914b, 1939) – had no doubt that the origins of the double-entry bookkeeping should be sought not in Genoa, Milan or Venice, but in Florence or in the books of Florentine merchants abroad.

In the middle of the 20th century there were renewed the archival research aiming to find the early traces of double-entry bookkeeping. In this vein we should emphasise the works of Littleton (1966), Yamey (1940, 1994), Zerbi (1952), Melis (1950, 1960, 1972), Martinelli (1974), Lee (1973, 1977), etc. The listed authors saw the existence of a double entry, starting from the two folios in the bankers' book of 1211. A double entry involves the registration of equal amounts characterising an economic transaction in the debit of one account and in the credit of another one. Thus, none of them ever wrote about the double-entry bookkeeping in the 13th century. The accounting systems of that time met only one challenge – the reflection of the settlement transactions between different debtors and creditors. They did not calculate the financial result and did not provide information on the flow of money and other property. This is particularly confirmed in the work of Sangster (2016), who states that the double entry originated in banks (but not in commerce), precisely in Florence, but nowhere else.

As can be seen from the above, the ledger of the G. Farolfi branch office (1299-1300) can be the only one to be recognised as the first preserved book of double-entry bookkeeping.

The first to describe the ledger of the Giovanni Farolfi's branch office was Melis in 1950. The researcher, even clearing that "he did not peruse all the long entries in the accounts of the company", came to the following conclusion: "The existence of the accounts with the sections "Debtor" and "Creditor" (the accounts were kept according to Tuscan rules – in a form of a paragraph), the application of the unified monetary estimates (*livre tournois* of France, divided into 20 *sols*, and each sol into 12 *deniers.*), the compulsory occurrence of the cross-references to the entries throw no doubt that the ledger was kept due to all the canons of double-entry bookkeeping" (Melis 1950, p. 485). In addition, he mentioned that "inside the frontispiece there was a wet ink: "The book of debtors and creditors..., kept by the Floretine rules" [ib.].

A decade later he dedicated a particular article to that accounting book (Melis 1960, pp. 347–356).

In 1972 Melis published the photocopy of the account Carte Strozziane serie II, 84 bis, c. 83R, in which, in his analysis, there was presented the "Fixtures (masserizie)" account (Melis 1972, pp. 384–385). The distinguished Italian scientist had no doubts that there was the account, which reflected the procedure of the depreciation provision. This point was disregarded by other researchers. At first, the authors shared the opinion of F. Melis. However, our latest discoveries let us state that in this certain case the accrual of depreciation was hardly probable. We believe there was not the "Fixtures (*masserizie*)" account being viewed, but the "Utensil" account. So, the preferable inference is that the described procedure could stand for losses on utensils breakage, its stealing and suchlike factors. As an utmost presumption, the authors approve of the utensils impairment, although there is no allusion of the inventory and the reassessment in the entries of the account.

Concerning the structure of the ledger F. Melis specified: "At the same time it is unfortunate that many of G. Farolfi's company accounts are missing: they begin from carta 47 and end up by carta 108; some of them are damaged, others have discoloured" (Melis 1950, p. 485).

In 1952 Castellani (1952, pp. 708–803), unlike F. Melis, examined in greater detail and translated into present-day Italian language each account of the book. Castellani was a philologist. He did not seek to delve into the specifics of the content of the accounting entries, but to preserve the book for the generations to come and for the future researchers. At the same time he was convinced the ledger was kept according to all the rules of double-entry bookkeeping.

Raymond de Roover mentioned the ledger in 1956. However, he did never avouch to have been working with the book of Farolfi's company: "Accounts for operating results also appear in another fragment published by Professor Castellani" (Roover 1956, p. 119).

He reckoned up the Farolfi accounting system in the following: "It may well be early example of double entry, since all the entries, save those relating to cash trans-actions, have cross-references to corresponding debits or credits. In the case of cash transactions, the absence of such references proves nothing, since receipts and dis-bursements were recorded in a separate book, called *libra dell'entrata e dell'uscita*, which was complementary to the ledger and served both as cashbook and as cash account. Merchandise accounts, it seems, were also kept separately in a *libra rosso* or red book. Of course, sectioning the ledger is perfectly compatible with double entry. It is even indicative of better organization, since it permits dividing the work among several book-keepers" [ib.].

Roover was especially explicit while describing the very well-know today exam-ples of deferred expenses, which occurred as a result of rental payments for the house or the store.

On a final note the American professor concluded: "Further improvement was achieved by placing the amounts in extension columns instead of inserting them in the narrative Summations were thereby greatly facilitated, but the use of Roman numerals continued to impose the aid of the abacus. It is true that a medieval clerk who knew "the lines", that is, the use of the abacus, could cast his counters as fast as we can operate today with adding or calculating machines" (Roover 1956, pp. 119–1200. Considering that Roover never saw the Farolfi ledger and limited himself by using Castellani's translation, he did not say anything about its structure or condition.

More rigorous research of this register was conducted by the American scientists Alvaro Martinelli and Geoffrey Alan Lee in the seventies of the past century. Yet, we cannot confirm whether they worked with the genuine source or were confined to the translation of professor Castellani.

A. Martinelli wrote: "This manuscript is kept in the State Archives of Florence, and it consists of 56 paper folios with original numbering from 48 to 110; folios 66, 74, 87, 105 and 106 are missing, and also missing are all the folios from 111 to 129, where several entries mentioned in the manuscript were located. In this ledger there is just one handwriting, which may be attributed to Matino (Amatino) Mannucci, partner of a company, which had its head office in Nimes. The ledger belonged to its branch-house of Salon, called Sallone by the Italians. The name of the company was probably "Giovanni Farolfi and partners", however quite often debts and credits were settled by ser Giovanni Filippi; there are also accounts of ser Giovanni Filippi and of one Pagno Franchi, which apparently are mixed with those of Giovanni Farolfi and partners, as if they constituted one account" (Martinelli 1974, p. 402).

Further he added: "Following a well-established practice, this ledger was divided into two sections: in the first section, from folio 1 to folio 92, there is record of all the accounts beginning with a debit entry, whereas in the second section, from folio 93 to folio 129, were recorded all the accounts beginning with a credit entry. Many of the accounts are personal, but there are also several impersonal accounts: rent for houses and shops, horses, wool, furniture and fixture, and so on" (Martinelli 1974, p. 403).

Lee claimed: "This document is preserved in Archivio di Stato, Florence, and filed under Carte Strozziane, 2a serie, n. 84 bis. It consists of 56 leaves of paper, about 33 × 24 cm., consequently numbered. The first 47 leaves are lost; the extant ones run from 48 to 110, with numbers 66, 74, 87, 93, 105, 106 and 109 missing. Leaves 92, 94 and 103 are damaged in their lower halves, and 92 and 110 are out of sequence – 92 bound after 96, and 110 at the beginning. In this article reference is made to the leaves, recto (r.) and verso (v.), as in Castellani. The bookkeeper, however, numbered only the recto of each pair of facing pages, and counted them as one folio; thus, a cross-reference to an entry *nel lxiii carte* meant that the counterpart appeared on either 62v, or 63r. Some references extend beyond the extant pages, mentioning folio numbers as high as 129, and it therefore seems that less than half the book has survived" (Lee 1977, pp. 79–80).

Our list of the missing pages somehow differs from the ones in the paper reviewed above. Summarizing the literature review we would like to pay attention to the fact that different researchers named the number of the first preserved page in various ways and were ambiguous in defining the numbers of the rest of the lost pages.

Purpose of the article. Virtual representation of the preserved pages in the ledger of the branch office in Salon of the company of Giovanni Farolfi in the form that was relevant at the end of 1300.

3 Statement of Basic Materials

The set of registers of the reporting period (1299–1300) consisted of five books – General ledger, two merchandise ledgers, expenses ledger, and cash book – constituted what looks very like a true double entry system, though for the Salon branch only. In addition, there were at least two subsidiary books, occasionally referred to the General ledger. The first one was the *libro piloso*, the second one – the "Memorandum ledger" (*quaderno memoriale*), for which no folio references are given. There are some references to these books in the General ledger.

As Lee wrote (which is also proved in our research): "The main merchandise accounts – for wheat, barley, oats, olive oil, wine, wool, and yarn – were contained in a separate "Red Book" (*libro rosso*). These accounts were debited with purchases (in the front of the ledger) and credited with sales (in the rear half), and closing inventory balances were transferred to the debit of the balance account in the General ledger" (Lee 1977, pp. 82).

The next goods register was the Cloth ledger (*quaderno dei panni*). This obviously contained merchandise accounts (as in the Red Book) for various kinds of cloth.

A great number of reference entries in the General ledger were made to the Expenses ledger (*quaderno delle spese*); there are references to folios from 6 to 37. Lee was confident that: "It seems to have contained accounts relative to fixtures (*masserizie*) and current expenses (*spese corse*), and possibly other item. Corresponding entries appear in the General ledger; then the closing totals of the Expenses Ledger accounts have been transferred to Fixtures, and Current Expenses, Accounts, and further entries made therein, as with the Cloth Ledger" (Lee 1977, p. 83). From our point of view, on page Carte Strozziane serie II, 84 bis, c. 83R there is presented the "Utensil" account, but not the "Fixtures (masserizie)" account, which is exactly the case when we disagree with F. Melis. The entries in the account for utensils say the following:

- Utensil **must give** on August 25, 300; what is recorded in the Book of Expenses, c. 7 95 *livrs, 5 sous*
- February 27, '99; it **must have** what is written in the Book of Expenses, on c. 8 *26 livrs, 17 sous, 2 deniers*
- The balance of 68 livrs 7 sous 10 deniers is defined in the first calende of September '99. Utensil for the certain sum *is given* to Giovanni Filippi, the sum of what he **has to give** is confirmed on c. 91 *68 livrs, 7 sous, 10 deniers*

As is evident from the foregoing, the Expenses ledger (*quaderno delle spese*) is being addressed twice, indeed.

The analysis of the entries in the General ledger rises up a very interesting issue. Why is there only one reference to the Cash Book? The answer to it is given by one of the most prestigious researchers of this book: "Last among the main books of account was the Cash Book (*libro dell'entrata e dell'escita*). This is never mentioned in the text, and no folio references to it are given for the numerous debits and credits relating to cash paid and received, or for the sum of 1/. 10 s. O d. transferred at 16 August 1300 to the debit of the balance account. This item is described as moneys, which we gave them in cash in Nimes: Matino counted it out." It must be the final cash balance, as it follows a series of debits for ending inventories of various goods, from the Red Book. The Cash Book was probably in the form then standard in Italy, with receipts in the front half, payments in the rear half, and periodical balancing of the two totals. The absence of folio references in the General Ledger is explained by the relative ease of referring back to the Cash Book by dates alone; there would, for obvious reasons, have been ledger references in the Cash Book" (Lee 1977, p. 84).

Disconcertingly, however, all the account books, except for the General ledger, were lost which makes it impossible to check on the back references.

The White ledger, which is the General ledger of the previous period of 1298, is also missing. The opening balances in many accounts of the existing General ledger have references to the closing balances in the White ledger. In all probability, there was not the opening balance in the General ledger. Balances were directly transferred from the account in the old ledger (White ledger) to the account in the new one (General ledger).

G. A. Lee expressed a high opinion of that accounting system: "All these records add up to a bookkeeping system of uncommon sophistication, even by contemporary Italian standards. Wonder grows when one remembers that these were the books of one branch only of a sizeable mercantile enterprise, and they arouse in the inquirer a strong, but fruitless, desire to know what elaborate arrangements must have been in force at the Nimes head office of the company" (Lee 1977, pp. 85–86).

At this point it is important to make a determination on the page numbering in the General ledger. The aforementioned authors use two concepts. The first of them suggests to accept the number assignment operated by A. Castellani. In this case, the right page of the double spread should be named as Recto (front side) while its opposite page should have the same number and be named as Verso. Another group of researchers align with the numbering offered by Matino Manucci: both of the pages of the double spread (folio) are assigned with one number. Yet, the adherents of such approach call the left side of the folio as "Verso".

The first concept appears to be improper to us as it supposes that the "Recto" and the "Verso" pages have different numbers. In our opinion, to comply with the idea of Matino Manucci, it is necessary to label both of the pages on the double spread with one number. In terms of our research we additionally put "R" ("Right") to the upper right corners of the ledger (the place on which the medieval accountant wrote the page number) and "L" ("Left") to the previous page of each double spread.

While investigating the information system of the Farolfi company, we managed to state that according to present knowledge the ledger consists of 58 numbered double spreads (folios). The first three of them are provided for archival reference, although quite possible that the right leaf of the second folio (which is in handwriting) can be referred to the late 13th century. Consequently, to the same time period can be referred the left leaf of the third double spread as there is an ink trace left after some recording on the second folio.

The accounting information begins on the right page of the folio 04 (archival numbering, Fig. 1).

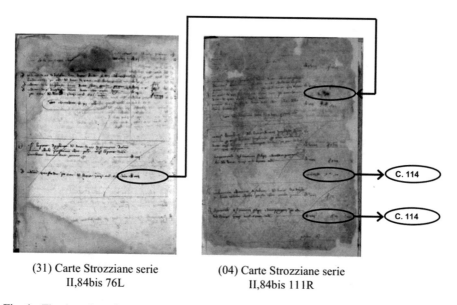

(31) Carte Strozziane serie
II,84bis 76L

(04) Carte Strozziane serie
II,84bis 111R

Fig. 1. The detection of the page number on the photocopy (04) (Here and below, reprinted with permission of the Ministry of heritage, culture and tourism of the Republic of Italy. Protocol №627/646.13.10 from 5 February 2014.)

As is seen from Fig. 1, the folio contains three accounts. All of them are related to the creditors' accounts, which should be apparently placed in the second part of the ledger. The information in the first account is partly missing. There is only the credit total of Lb. 18 s. 8 remained. Further, three debit entries follow. The second one has an explicit reference to carta 76. The second account on the left side of folio 76 pertains to the debtor Messer the Prior of Aghiera and contains two records:

Messer the Prior of Aghiera **must give**, on June 30th 1300, for we loaned him cash. Messer the Prior took them and Bacchera renewed. He must give wheat for… lb. 4.

He **gave**, on this day, and we put where he **must have** here at c. 111, lb. 4.

The second credit entry is of high importance because it addresses to the studied page and identifies it as 111.

Such scenario has fazed us a bit as this discovery contradicts to all known researches. More importantly, if the page of interest is admitted to be 111 R (Right), then the set of preserved pages should keep the page 112 L (Left). But it is the one that does not exist.

Another two accounts on page 111 R (according to a cross-reference) pertains to the creditors Giames Brocchiere and Giacomino key-maker of Sallone. Both the accounts address to folio 114.

As further studies show, about a dozen of the preserved creditors' accounts contain a reference to folio 114. This allows us to assume that on one of the sides (or probably on both of them) there was the final page of "Creditors" of the Trial Balance.

Hereafter we identified the second folio in the folder of the preserved pages of the ledger (Carte Strozziane serie II,84bis_05). There is the translation of the left page of the folio (05) presented below:

Lippo Baldovini **must have** the […] of August year 300 lb. 24 (tournois) for 45 *somate di civada* we bought from him at s. 10 d. 5 for each *somata*; we put at purchased merchandise in the Red Book at […], lb. 24.

We **gave** the 6 of August, the same year, lb. 24 to.; we put where it **must give** here at c. 82, lb. 24.

"*Spese corse*" (Expenses occurred) **must have**, the 15 of July 99 lb. 49 s. 19 d. 1 to.; we put where they **had to give** at the Expenses Notebook at c. 37, lb. 49 s. 19 here at c. 87".

Further, the translation of the first page is presented (Table 1).

The page represents two accounts. The first one pertains to the creditor Lippo Baldovin and addresses to carta 82.

Figure 2 shows a diagram of the real accounts photocopies, recordings of which correspond to the accounts placed on the folio of interest. The offset entry in the first (upper) account on page 82R has a sum of lb. 24 and addresses to carta 110.

The same true is for the second account on this folio. It directly refers to page 87. Indeed, the account on page 87R reflects the sum of lb. 49 s. 19 d. 1 and a reference back to 110. The left page on folio (05) in the set of photocopies, prepared by the State archive of Florence, is free from any other references. It provides basis for identification of the page number as 110 L.

Unfortunately, the set of the preserved pages of the Farolfi company ledger is rated as one of the archival artefacts, which cannot be checked out to the researches, but can be shared in the form of photocopies. Before identifying the page number on folio (04) we had an idea that the mentioned page was the front side in regard to the opposite side 110 L. Consequently, we assigned it number 109 R. Such an approach logically matched our system of virtual computer reconstruction of the preserved ledger pages according to the rules set by Matino Manucci in his day.

Our next task is concerned with the identification of the page number of the first preserved page in the General ledger placed on the right side of folio (05).

Table 1. The translation of folio 04

....	....
Not readable	Not readable
...	Lb. 18 s. 8
We *gave*, on August 8th of the aforementioned year, we *gave* for him to Giames Brocchiere for a mule bought from him by messer the Prior. He *must have* here below	Lb. 9 s. 17
We gave on June 30th of this year. We put where he should give here at c. 76	Lb. 4
We *gave* on August 11th, in cash. Addo took them, Bacchera renewed	Lb. 4 s. 11
The sum we *gave* is lb. 18 s. 8	
Giames Brocchiere *must have* on August 8th 1300. They are for messer the Prior of Aghiera for a mule that messer the Prior bought from him. We put where he *must have*, here above	Lb. 9 s. 7
We assign it to ser Giovanni Filippi and Co. They *must have* here at c. 114	Lb. 9 s. 17
Giacomino key-maker of Sallone *must have* on August 10th 1300, he lent cash to us, Perone brought them	Lb. 8 s. 4
We assign it to ser Giovanni Filippi and Co. They *must have* here at c. 114	Lb. 8 s. 3 d. 4

Table 2 presents the translation of the aforementioned page.

The opening balance of the debtor's account The Archbishop of Arli was transferred from page 52 of the White Book, the General ledger of the previous accounting period, which ended up in 1298. The debit (upper) section of the account consists of six entries and is completed with the total equalling to lb. 268 s. 12 d. 5 tor. Two of the entries contain cross-references to the pages of this ledger, which are carta 60 (the sum of lb. 10) and carta 76 (the sum of lb. 3 s. 6) that is clearly illustrated in Fig. 2. Therefore, the cross-references in the mentioned accounts address to page 48, which lets us identify the right side of the folio as page 48 R.

The credit section of the debtor's account on page 48 R (the bottom of the account) includes four entries totalling to lb. 268 s. 12 d. 5 tor. Two of them have references to the pages of the General ledger: to carta 68 (the sum of lb. 11 s. 10) and to carta 74 (the sum of lb. 113 s. 4 d. 5). It is impossible to verify the correctness of the entries, as the leaves with these pages were lost.

Thus, there is no doubt that the first preserved page of the Farolfi ledger is 48 R.

Further, the researchers ensured that all the folios of the preserved pages in the General ledger, offered by the archive, starting with the folio (06) (50 L–50 R) up to the folio 22 (65 L–65 R) do not relate to the lost pages. However, there is a number written by the medieval accountant in the upper right corner of the right page which is 67. So, we have made an attempt to identify the true number of the left page on folio (23).

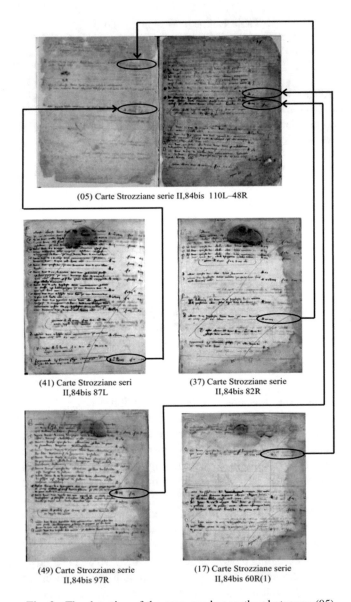

Fig. 2. The detection of the page number on the photocopy (05)

We have made the translation of the first three records in the account:

Our cloths *must give*, at the [...] of June thirteen hundred, lbs. 31 s. 4 tornesi ... our partners *must have* in the White book at [...] with other money; we took them from the book of cloths, at c. 16.

They *gave* at the 15th of October of the same year, lbs. 20 s. 1 d. 9 for 44 and one half *channe* and one half *palmo* which we sold to several persons, as appears in this book in the fourth following ...

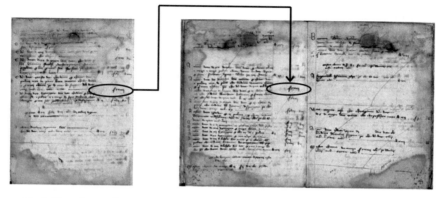

Carte Strozziane serie   **Carte Strozziane serie II,84bis_23 66L-67R**
II,84bi c. 58 L

Fig. 3. The detection of the page number on the photocopy (23)

Table 2. The translation of the right page on folio 05

The Archbishop of Arli **must give**, the 9 of November, year 99, lb. 174 s. 12 d. 5. tor; we put where it had to give at the White Notebook at c. 52	lb. 174 s. 12 [...]
And he **must give**, the 30 of March, year 300, s. 9 tor. M for 1 mark of *poggiesi* that Oriache his equerry.	s. 9
And he **must give** the 29 of April, said year, lb. 30 tor.; we loaned them in cash, Guillielmo his *somelliere* took; [...] Bacchera.	lb. 30
And he **must give**, the 13 of May, said year, lb. 1 tor.; he had f. 52 of gold and s. 12 tor. For s. 19 each; mr. Guillielmo d'Ooriache took. [...] Bacchera and Matino. Sum:	b. 50
And he **must give**, the 10 of April, said year, lb. 10 to.; we **gave** for him to Bertrano de le Fascia; we put where Giomino keymaker of Sallone **should give** before at c. 60	lb. 10
And he **must give**, the 10 of June, said year, lb. 3 s. 6 tor.; for a piece of black *zendado* sent by Giovanni Farolfi and our Co. in Nimmisi; we put for they **must have** at c. 97	lb. 3 s. 6
Total sum, he **must give** lb. 268 s. 12 d. 5 tor; count made the 11 of January year 99	
They **gave**, the 16 of May said year, lb. lb. 99 s. 18, which Matino had in Arli, cash	b. 99 s. 18
They **gave**, the 10 of June, said year, lb. 11 s. 10, which dr. Tommasino Farolfi had from mr. Guiglielmo di montauto; we put because Tommasino **must give** at c. 68	lb. 11 s. 10
They **gave**, this day, lb. 15, because they **must have** from us for collecting of the grain of San Mieri and San Ciemaso.	lb. 15
They **gave**, the 16 of September, year 99, lb. 113 s. 4 d. 5; we put as he **must give** at c. 74	lb. 113 s. 4 d. 5
Total sum, lb. 268 s. 12 d. 5 tor.	

They **gave** at the 10th of May thirteen hundred, s. 19 for two *channe* and *one nalmo* of cloth which we sold to Matino Mannuci for s. 9 per *channa*: we posted that he **must give** here at c. 58.

The first entry in the debit of the "Cloth" account testifies that its opening balance in the sum of lbs. 31 s. 4 tornesi was transferred from the book of cloths to carta 16. The record, which arouse our interest, is the third one as there definitely is the page number (58) and the sum (s. 19). In the account of the left side of folio (58) there is a corresponding entry with the mentioned sum and the reference to carta 66. Consequently, folio (23) presents pages 66 L and 67 R, when the double spread 66 Right – 67 Left is missing.

The lost leaves 74 Right – 75 Left; 87 Right – 88 Left; 93 Right – 94 Left were detected in a similar manner.

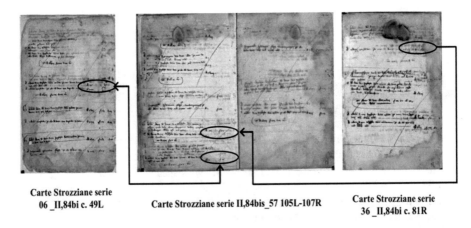

Carte Strozziane serie
06 _II,84bi c. 49L   Carte Strozziane serie II,84bis_57 105L-107R   Carte Strozziane serie
36 _II,84bi c. 81R

Fig. 4. The detection of the page number on the photocopy (57)

Some difficulties appeared while we tried to identify the page on folio 57 (Fig. 4). The medieval accountant assigned the number 107 to the right page. However, none of the references can be verified as they are being addressed either to the pages 113 or 114 of the General ledger, which are missing, or to the page 19 of the Book of Expenses. Thus, there is no choice left but to agree with our colleague.

The left page on folio 57 contains two cross-references in the third account: to the first account on page 49 L (the sum of lb. 113 s. 4) and to the first account on page 81R (the sum of lb. 116).

The given arguments let us state that placed pages on folio 57 are 105 L and 107 R. It gives us a thought that there was not one leaf lost but two whole: 105 Right – 106 Left and 106 Right – 107 Left (Fig. 4).

Starting with the page 93 R and on folio 114 (the right and the left pages cannot be identified) there were placed the creditors' accounts. Besides that, we managed to find about a dozen of references to the page 114 which are related to the transfer of the creditors' accounts balances. So, the reproduction of the page 114 is predetermined. Yet, it is another line of research.

The General ledger addresses to the Profits Account *(avanzi)*, which was on the last pages (127–129) of the General ledger. Concerning the Losses Account, it probably occupied several pages in the Book of Expenses.

Table 3 presents a refined list of pages that were affected by some changes. This way it is easier to depict the pages that were lost both according to the generally accepted page numeration and to the original numbering used by the medieval book-keeper. It is clearly shown that the pages from the end of the book are filed to the beginning of the General ledger and vice versa. Some pages from the middle of the book are located at the end of the register.

Table 3. The results of the study on the arrangement of the preserved pages in the ledger of the branch office of the company of Giovanni Farolfi in Salon (1299–1300)

№№ photo	Preserved pages				Missing leaves	
	Generally accepted page numeration		Numbering by Manucci		Generally accepted page numeration	Numbering by Manucci
	Back side (V)	Front side (R)	Left (L)	Right (R)	(R/V)	(R/L)
–	–	–	–	–	2R–47V	2R–48L
04	–	109R	–	109R		
05	109V	48R	110L	48R		
06	48V	49R	49L	49R		
…	…	…	…	…		
23	65V	67R	66L	67R		
–	–	–	–	–	66R–66V	66R–67L
24	67V	68R	68L	68R		
…	…	…	…	…		
29	72V	73R	73L	73R		
–	–	–	–	–	74R–74V	74R–75L
30	73V	75R	74L	75R		
…	…	…	…	…		
42	86V	88R	87L	88R		
–	–	–	–	–	87R–87V	87R–88L
43	88V	89R	89L	89R		
…	…	…	…	…		
46	91V	94R	92L	94R		
–	–	–	–	–	92R–92V	92R–93L
–	–	–	–	–	93R–93V	93R–94L
47	94V	95R	93L	95R		
…	…	…	…	…		
56	103V	104R	104L	104R		
–	–	–	–	–	105R–105V	105R–106L
–	–	–	–	–	106R–106V	106R–107L

(continued)

Table 3. (*continued*)

№№ photo	Preserved pages				Missing leaves	
	Generally accepted page numeration		Numbering by Manucci		Generally accepted page numeration	Numbering by Manucci
	Back side (V)	Front side (R)	Left (L)	Right (R)	(R/V)	(R/L)
57	104V	107R	105L	107R		
58	107 V	108R	108L	108R		
59	108 V	92R	109L	92R		
60	92 V	–	93L	–		
–	–	–	–	–	110R–129(x)	110R–129(x)

The implemented research allows us to rearrange the preserved pages in the General ledger into a sequence set by Matino Manucci seven centuries ago. The publication limited in volume does not let us describe the work accomplished in details. At the same time it is obvious that it provides basis for a further more profound research of the precious archival materials. In the number of key studies there should be paved the way for the in-depth studies of the medieval accounting procedures and in the first place for the preparation procedure of the final (summary) pages of the sections "Debtors" and "Creditors" in the Trial balance. Concerning the final credit page which was lost, we can say that there were not mentioned any attempts to reconstruct it virtually before.

4   Conclusion

Summarising the results of the research, it is necessary to note the importance of the modern information technologies application in conducting of the researches in any areas of science. Researchers frequently face the issues of historical documents preservation, so here the digitisation comes to the aid helping not only to preserve the relics for the future generations, but also to conduct the researches without regular attendance of the archives. The availability of the documents in the free access is an invaluable support for any historian.

Nevertheless, the fact remains of no small interest that the modern technologies can allow not only to save, but to make the field of the research more informative as well. As is the case with the General ledger of the Farolfi company the modern information technologies provide both the reproduction of the lost information and the revival of the historical document in its original form to the extent possible.

As a result of the research, each preserved page of the book was identified, their historical numbering was restored, and the sequence that was established seven centuries ago was recreated. This will allow one to carry out the in-depth studies of the first preserved ledger of the double-entry bookkeeping application. In particular, the authors expect to obtain evidence of a two-stage reconstruction of the "Debtors" section in the Trial Balance, when the balances of the unclosed accounts were accumulated on separate pages at first, and then the results from the pages were transferred to a summary table. In addition, there will be made an attempt to reconstruct an irrevocably lost page 114, on which the balances of the unclosed pages of the "Creditors" section were collected.

References

Alfieri, V.: La partita doppia applicata alle scritture delle antiche aziende mercantili veneziane/ V. Alfieri. Roma (1911)

Besta, F.: La Ragioneria, 2nd Edn. vol. 3. Rirea, Rome (1909). Facsimile Reprint 2007

Besta, F.: La ragioneria, seconda edizione riveduta ed ampliata col concorso dei professori Vittorio Alfieri, Carlo Ghidiglia, Pietro Rigobon. Parte prima. vol. III. Vallardi, Milano (1922)

Castellani, A.: Nuovi testi fiorentini del Dugento e dei primi del Trecento. T. 2. The ledger of Giovanni Farolfi & Company, 1299—1300, is transcribed, almost entire, Firenze (1952)

Ceccherelli, A.: I libri di mercatura della Banca Medici e l'applicazione della partita doppia a Firenze nel secolo decimoquarto. Bemporad, Firenze (1913)

Ceccherelli, A.: Le funzioni contabili e giuridiche del bilancio delle società medievali. Rivista Italiana di Ragioneria **14**(8), 371–378 (1914a)

Ceccherelli, A.: Le funzioni contabili e giuridiche del bilancio delle società medievali (Continuazione e fine). Rivista Italiana di Ragioneria **14**(10), 436–44 (1914b)

Ceccherelli, A.: l linguaggio dei bilanci. Formazione e interpretazione dei bilanci commerciali. Le Monnier, Firenze (1939)

De Roover, R.: The development of accounting prior to Luca Pacioli according to the account-books of Medieval merchants. In: Littleton, A.C., Yamey, B.S. (eds.) Studies in the History of Accounting. pp. 114–174. London (1956)

Kuter, M.I., Gurskaya, M.M., Sidiropulo, O.A.: The Genoese commune massari's ledger of 1340: the first computer modeling experience and its results. J. Mod. Account. Audit. **9**(2), 212–229 (2013). ISSN 1548–6583

Lee, G.A.: The Development of Italian bookkeeping 1211–1300. Abacus **9**(2), 137–155 (1973)

Lee, G.A.: The coming of age of double entry: the giovanni farolfi ledger of 1299–1300. Account. Hist. J. **4**(2), 79–96 (1977)

Littleton, A.C.: Accounting evolution to 1900/A.C. Littleton. Russel & Russel, New York (1966)

Martinelli, A.: The origination and evolution of double-entry of bookkeeping to 1440. ProQuest Dissertations and Theses, 1974, ProQuest Dissertations & Theses Global pg. n/a.н (1974)

Melis, F.: Storia della ragioneria. Zuffi, Bologna (1950)

Melis, F.: Osservazioni preparatorie al bilancio nei conti della Compagnia Farolfi, nel 1300, in Studi di Ragioneria e Tecnica economica: Scritti in onore del Prof. Alberto Ceccherelli. Pubblicazioni dell'Università degli Studi di Firenze, Firenze, pp. 347–356 (1960)

Melis, F.: Documenti per la storia economica dei secoli XIII–XVI. Firenze (1972)

Rigobon, P.: Cenni sulla Contabilita delle Antiche Corporazioni Religiose in Toscana, Ragioniere, VII (Sez. II), pp. 13–24 (1891a)

Rigobon, P.: Alcuni appunti storico bibliografici etc. « Monografie edite in onore di F. Besta » . Milano (s. a.) (1891b)

Sangster, A.: The genesis of double entry bookkeeping. Account. Rev. **91**(1), 299–315 (2016)

Yamey, B.S.: The Functional Development of Double-Entry Bookkeeping, Accounting Research Association, Bulletin No. 7 (1940)

Yamey, B.S.: Luca Pacioli. Exposition of double entry bookkeeping. (Venice)/Albrizzi editore, Venice (1994)

Zerbi, T.: Le Origini della partita dopia: Gestioni aziendali e situazioni di mercato nei secoli XIV e XV (Milan: Marzorati) (1952)

The Analysis of the Final "Debtors" Section Page Preparation in the Trial Balance of Giovanni Farolfi's Company Branch Office

Marina Gurskaya(✉) , Mikhail Kuter ,
and Angelina Andreenkova

Kuban State University, 149 Stavropolskaya Street, 350040 Krasnodar, Russia
marinagurskaya@mail.ru

Abstract. In the second half of the last century the General ledger of the branch office of Giovanni Farolfi's company in Salon was recognized as being the earliest preserved example of double-entry bookkeeping application. This archival source is mentioned in many historical papers, however the number of real researchers of the accounting system of the last 13th century is limited. The paper presents the study results of the study on the accounting procedure of the "Debtors" section preparation in the Trial balance. It also proves the existence of the two-stage procedure, which supposes the preparation of the intermediate analytical pages and their following summarizing.

Keywords: The Farolfi company · Double entry · Trial balance
Intermediate tables · Summary page

1 Introduction

The General Ledger of the branch office in Salon (1299–1300), which belonged to a Tuscan merchant, Giovanni Farolfi is recognized as the earliest preserved example of double-entry bookkeeping application. A great number of articles describing this unique historical memorial of the accounting development were published, however the limited number of researchers had really worked with the book.

The present study focuses on the procedure of the "Debtors" section preparation in the Trial balance. The applied methodology is peculiar only to the Russian school of historical research – the method of cross-references and logical-analytical reconstruction. Therefore, the paper provides convincing argumentation, confirming the two-stage nature of the procedure of the Trial balance preparation, which was used seven centuries ago.

Thus, the given article aims to identify the elements that are characteristic of the accounting system of interest on the basis of the Trial Balance procedure modelling.

2 Review of Prior Literature

The General ledger of the branch office of Giovanni Farolfi's company was described by a limited number of authors: Melis (1950, p. 485; 1960, pp. 347–356; 1972, pp. 384–385), de Roover (1956, p. 119), Martinelli (1974, pp. 402–422), Lee (1977, pp. 79–95), Smith (2008, pp. 143–156).

© Springer International Publishing AG, part of Springer Nature 2018
T. Antipova and Á. Rocha (Eds.): MosITS 2017, AISC 724, pp. 185–194, 2018.
https://doi.org/10.1007/978-3-319-74980-8_17

Seems like Melis was the one who was studying the preserved pages of the book firsthand, but his research did not touch upon the issues of our interest. Besides, the distinguished researcher of the Italian archives confessed "he did not peruse all the long entries in the accounts of the company".

Furthermore, we cannot fail to emphasize Professor Castellani (1952, pp. 708–803), who, being a language expert, translated the whole ledger to modern Italian language for a noble cause – to preserve the medieval text for the future generations. Castellani was sure the ledger was kept according to all the rules of double-entry bookkeeping.

The ledger description by de Roover was limited by one page in a chapter of a multi-authored monograph of 61 pages in general (De Roover 1956, pp. 114–174), at which point he did not see the archival artefact, but was making reference to the work of A. Castellani.

The research of G.A. Lee can be characterized as a profound one. On the issue of concern he wrote the following: "Finally comes the acid test of any alleged double entry system – the balancing process. Amatino Manucci does not disappoint. On page 88r. begins an account headed *Giovanni Farolfi e' compangni n(ost)ri di Nimmisi*. To its debit side are transferred almost all the debit balances open on the General Ledger to that point, including some on missing folios (2 to 7) at the beginning, after which there is an abrupt jump to the extant part of the ledger, from folio 48 onwards. The account continues to the bottom of page 90v., with each page separately totaled; on page 91r. comes a summary of open debit balances on the Red Book—ending inventories of produce – followed by what is clearly the ending balance of cash in hand, from the Cash Book. (Balances on the Cloth Ledger and Expenses Ledger have already been taken up into the General Ledger). Page 91v. sweeps up a few debit balances previously overlooked, or added after the balancing began, then summarizes the totals of pages 88r to 91r, making a grand total of 2,745 *l.* 6 *s.* 1 *d.* tournois" (Lee 1977, p. 91).

In the meantime the prestigious scientist, in all probability, did not work with the ledger itself, but based his logical findings while replicating the General ledger of Farolfi by means of Castellani's translation. Apparently, for that reason Lee did not assign the real numbers in the ledger, the ones used by the medieval accountant, but confined himself to the numbers given by the translator, A. Castellani.

Smith (2008, pp. 143–156) appears to have worked neither with the genuine ledger nor with Castellani's translation. Essentially, she used to align with the work of her mentor. Her paper is focused on the comparison of the contribution of Luca Paciolli and Amatino Manucci to the development of accounting. Therefore, it shows the role of Geoffrey Alan Lee in the modern science of accounting history education.

The following statement of A. Martinelli motived us to begin our research: "The section of the ledger from folio 88 to folio 92 contains credits which had been assegnati, that is credits which had to be paid to the head office in Nimes. The usual expression was: We credited it to ser Giovanni Filippi; we posted that they must give further on to folio ... lbs" (Martinelli 1974, p. 404).

Such an opinion totally contradicts the aforementioned quotation of G.A. Lee. From our point of view, this is one of the few mistakes made in the work of Martinelli, the volume of which is almost 1000 pages.

The purpose of the article. Modelling the procedure of the final page of real, nominal and personal accounts of the "Debtors" section, proving the existence of these two stages of the mentioned procedure.

3   Statement of Basic Materials

The small number of researches of the earliest double-entry accounting ledger and the difference in their opinions concerning the eight accounts of the General Ledger, which preceded the final page of the "Debtors" section in the Trial balance

As a rule, while working with the medieval accounting complexes that are being kept in the archives of Italy, we use two main methods:

– the method of cross-references;
– the method of logical-analytical reconstruction.

The method of cross-references allows detecting or checking the page numbers in the accounting registers with the help of the records that contain the remained references on the examined page. The analysis of the references in the corresponding entries lets us identify the page of interest.

The method of logical and analytical reconstruction (MLAR) is intended to bind all the transactions, registered in all the journals (accounting books) that are included into the archival accounting complex, into one diagram. Its basis is the principle of decomposition of the indicators (vertically down), reflected in the Trial balance (Balance sheet) and in the "Profits and Losses" account. There used the principle of composition (aggregation) of the indicators (vertically up) in order to study the early stages of accounting (before the use of Trial balance (Balance sheet) and "Profits and Losses" account).

Such approach allows following up the formation mechanism of each primary, intermediate and resultant indicator. There also appears a possibility to identify the common patterns and the striking features of their calculation. The analysis of the accounting system and its comparison with the other accounting systems by time, areas and types of ownership are aimed to state the factors that contributed to the accounting establishment and development.

The detection of each page of the ledger and their virtual reconstruction into a sequence set by Amatino Manucci seven centuries ago could let us create a diagram of the final page of the "Debtors" section of the Trial balance (c. 92 L), Fig. 1. The translation of the summary page is presented in Table 1.

The first total was transferred from the page 86R, the last page, on which the debtors' accounts were placed. Therefore, it is necessary to pay attention to the fact that from the second page to the mentioned one there were jumbled up personal, real and nominal accounts.

On the aforementioned page one can find two accounts. The translation of the first of them is presented in Table 2. The second record in the account testifies about the increase in accounts receivable of Messer the archbishop of Arles in sum of Lb. 24 s. 13 d. 6 *to.* Concerning that the given sum was charged to account on carta

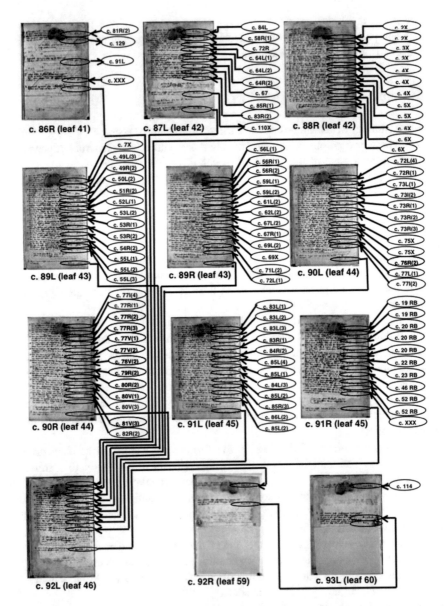

Fig. 1. The preparation of the summary page of the "Debtors" section in the Trial balance, the branch office in the company of G. Farolfi in Salon (1300)

129, on which the account for profits was placed, it can be assumed that this transaction is mean to be the reflection interest payments.

The second account on this page pertained to Rascasso, Jew of Sallone (the translation is shown in Table 3). Presumably, the account was the last one to be closed and its balance Lb. 95 s. 11 d. 8 *to.* was directly transferred to the summary page 92 L.

Table 1. The translation of Carte Strozziane serie II, 84bis_46 92 verso

1300	
And they *must give*, the […] of May, we *give* them in *Guagi*, as it can be seen at c. 86, we put where they *had to give*	lb. 95 s. 11 d. 8
And they *must give*, the 12 of May, year 300, we *give* to them for *spese corse* (expenses occurred), as it can be seen here at c. 87 in sum; we put where they *had to give* in said place	lb. 574 s. 5 d. 5
And they *must give*, the 7 of June, year 300, we put where they *had to give* here at c. 88	lb. 78 s. 6 d. 10
And they *must give*, the 24 of July, said year, we put where they *had to give* here at c. 89	lb. 203 s. 16
And they *must give*, the 4 of October, said year, we put where they *had to give* here, the same place of the other page	lb. 218 s. 8 d. 4
And they *must give*, the 10 of May, said year, we put where they *had to give* here at c. 90	lb. 81 s. 14 d. 3
And they *must give*, the 22 of June, said year, we put where they *had to give* here, at the same place of the other page	lb. 172 s. 16 d. 3
And they *must give*, the 20 of June, said year, we put where they *had to give* here at c. 91	lb 479 s. 3 d. 9
And they *must give*, the 15 of July, said year, we put where they *had to give* here at the same place of the other page	lb. 841 s. 3 d. 6
Total sum, they *must give* lb. 2745 s. 6 d. 1 tor., counted the 27 of June, year 300	
They *gave*, the 27 of June, said year, we put they *must give* here at the other page face	lb. 2745 s. 6 d. 1

Table 2. The translation of Carte Strozziane serie II,84bis_41 86 R(1)

1300	
Messer the archbishop of Arles *must give*, on the 25 of July 1299, we put where he should give here at c. 81	Lb. 151 s. 14 d. 5 *tor.*
And he *must give*, on calends of September 1300, for gift until this day, at d. 4 for one lb. we put in the account *avanzi* at c. 129	Lb. 24 s. 13 d. 6 *to.*
The total sum is lb. 176 s. 7 d. 11 *tor.* on calends of September 1300	
We put this sum to the account of ser Giovanni Filippi. We put that they *must give* at c. 91	Lb. 176 s. 7 d. 11 *to.*

Table 3. The translation of Carte Strozziane serie II,84bis_41 86 R(2)

1300	
Rascasso, Jew of Sallone, *must give* on the 15 of May 1300. He had belts, cups and garlands and many other things. He had them many times, Rascasso himself took them […]	Lb. 95 s. 11 d. 8 *to.*
And we put this sum to the account of ser Giovanni Filippi and Co. We put that they *must give* at c. 92	Lb. 95 s. 11 d. 8 *to.*

We should note that this is the only case in which not the intermediate total of the analytical page was transferred to the summary page, but the particular balance of the personal account.

The second total on the summary page: «And they **must give**, the 12 of May, year 300, we **give** to them for *spese corse* (expenses occurred), as it can be seen here at c. 87». As Table 4, in which all the expenses are gathered, shows, the accumulated total equalled lb. 624 s. 4 d. 1 *tor.* to the end of the accounting period. On page 101 the second account reflects two entries:

- "Spese corse" (Expenses occurred) **must have**, the 15 of July 99 lb. 49 s. 19 d. 1 to.; we put where they **had to give** at the Expenses Notebook at c. 37;
- We **gave** the 15 of July, said year, lb. 49 s. 19 d. 1 to.; we put where must give here at c. 87.

Table 4. The translation of Carte Strozziane serie II,84bis_42 87 L

1300	
Expenses **must give** on the 16 of April 1300. We put where they **should give** here at c. 84	Lb. 518 s. 15 d. 3 to.
And they **must give** on calends of July 1299, which we gave to Bettino Bruni our footman for … month that he lived with us in Sallone. We put where he **should give** here at c. 58	Lb. 1 s. 14 d. 9 to.
And they **must give** on the 10 of May of this year 1300, for what Tommasino's people expended a tour house in Sellone for food and beverage. We put where he **should give** here at c. 72	s. 9 d. 6 to.
And they **must give** on the 1 of March 1299. We put where he **should give** here at c. 64	Lb. 5 s. 14. 6 to.
And they **must give**, on this day. We put where they **should give** here in the said account	Lb. 10 s. 10 d. 8 to.
And they **must give** on the 15 of January of that year for Giovanni Farolfi and Co. We put where he **should give** here at c. 64. The money was for *iscotti* coming and going and for money exchange	Lb. 8 s. 14 d. 11 to.
And they **must give** on the 26 of August 1299, for Giovanni Farolfi and Co. We put where he **should give** here at c. 67	Lb. 5 s. 17 d. 11 to.
And they **must give**, on the 15 of July of this year 1300, We put where he **should give** here at c. 86	Lb. 29 s. 2 d. 2 to
And they **must give** on the 20 of January of 1299, for food and beverage from the of May 1299 to the 12 of August 1300. We put where he **should give** here at c. 83	Lb. 44 s. 4 d. 10 to.
The sum is lb. 624 s. 4 d. 1 *tor.*, on this day 14 of April 1300	
They **gave** the 16 of July 1299. We put where he **should give** here at c. 110	Lb. 49 s. 19 d. 1 to.
It remains lb. 574 s. 5 d. 5 *tor.*, on this day 12 of May 1300	
We **give** this balance to ser Giovanni Filippo end Co. We put where he **should give** here at c. 92, lb. 574 s. 5 d. 5	

This means that a part of the accumulated costs totalling lb. 49 s. 19 d. 1 to. was written off and carried forward to the account for losses in the Red Book on carta 37. Consequently, the sum of the accumulated (unallocated) costs in the analytical table (c. 87 L) was reduced by this amount and the capitalized balance was transferred to the summary table on page 92 L.

Some special attention can be drawn by the fact that there are no cross-references to the page numbers preceding 48 R, i.e. the first preserved page. Yet, if we analyse the second analytical page of the Trial balance (88 R, translation in Table 5), we can find out that the accounts balances, prepared on this page, were transferred from the folios 2–6.

The next analytical page of the Trial balance (89 L) addresses to folio 7, then there follow the references that can be checked: the third account on page 49 L (sum of lb. 17 s. 16 d. 4); the second account on page 49 R (sum of lb. 3 s. 18); the third account on page 50 L (sum of lb. 78 s. 17 d. 7); the second account on page 51 R (sum of lb. 56 s. 10 d. 6), etc.

Table 5. The translation of Carte Strozziane serie II,84bis_42 c. 88 R

1300	
Giovanni Farolfi and our partners of *Nimmisi* (Nimes) **must give**, on the Day of Ascension 1300, 20 of May, for 3 quintals of lamb's wool we assigned to them, we **must have** from Ponzo di Villa, valued at s. 22 for one quintal. We put where he **must give** here at c. 2	Lb. 4 s. 16 *to.*
And they **must give**, on this day, for 4 quintals of lambswool we assigned to them, we **must have** from Ferriere Bonsegnore, jew, valued s. 32 for one quintal. We put where he **must give** here at c. 2	Lb. 6 s. 8 *to.*
And they **must give**, on this day, for 10 quintals of lambswool we assigned to them, we **must have** from Astrughetto son of Astrugo and da Segnoretto son of Astrugo di Bornuolo, valued s. 32 for one quintal. We put where he **must give** here at c. 3	Lb. 16 *to.*
And they **must give**, on Pentecost 1300, in calends of June, for 13 *somate* wheat we assigned to them, we **must have** form Raimondo Milo and messer Ramondo Rostagni, each *sommata* was valued s. 22 *tor.* We put where he **must give** here at c. 3	Lb. 14 s. 6 *to.*
And they **must give**, on the Ascension of this year, 20 of May, for 1 quintal of lamb's wool and 1 of wool we assigned to them, we **must have** form Astrugo Durante. Lambswool was valued s. 32 for one quintal and the wool s. 22. We put where he **must give** here at c. 4	Lb. 2 s. 14 *to.*
And they **must give**, on this day, for 2 *uintals* of wool we assigned to them, we **must have** from Ramondo di Sinisernis, valued s. 22 for one quintal. We put where he **must give** here at c. 4	Lb. 2 s. 5I *to.*
And they **must give** on this day, for 12 quintals of wool we assigned to them, we **must have** form Tano da figliano, valued s. 22 for one quintal. We put where he **must give** here at c. 4	Lb. 13 s. 4 *to.*

(*continued*)

Table 5. (*continued*)

1300	
And they ***must give***, on St. John's day, the 24 of June, for 5 *somate* of wheat we assigned to them, we ***must have*** from Ugo Rugiastelli, valued s. 22 for one *sommata*. We put where he ***must give*** here at c. 5	Lb. 5 s. 10 *to.*
And they ***must give***, on this day, for one *sommata* of wheat we assigned to them, we should have from Pere Fustiere, valued at s. 22. We put where he ***must give*** here at c. 5	Lb. 1 s. 2 *to.*
And they ***must give***, this day, for 2 *somate* of burley we assigned to them, we ***must have*** form Pere di Montilzi, valued at s. 12 for one *sommata*. We put where he ***must give*** here at c. 6	Lb. 1 s. 4 *to.*
And they ***must give***, on St. Lawrence's day, the 10 of August, b. 3 *to* for 5 *somate* of barley we assigned to them, we ***must have*** from ser Guillielmo Marco, valued s. 12 for one *sommata*. We put where he ***must give*** here at c. 6	Lb. 3 *to.*
And they ***must give***, on this day, for VI *somate* and one *imina* of wheat nad II *somate* of barley we assigned to them, we ***must have*** form messer Alfante, valued the *somata* of wheat s. 22 and the barley s. 12. We put where he ***must give*** here at c. 6	Lb. 7 s. 18 d. 10 *to.*
The sum is lb. 78 s. 6 d. 10 *tor.*, on the 7 of June 1300	
They ***gave*** on the VII of June 1300 lb. 78 s. 6 d. 10 *tor.* We put where he ***must give*** here at c. 92	

All the balances of the closed debtors' accounts (real, nominal and personal) were transferred to eight analytical pages (87 L, 88 R–91 R; folio 87 R–88 L is missing, but it was not used in the preparation of the debit total of the Trial Balance). So, eight totals from each analytical page were transferred to the summary page (92 L), which was meant to prepare the total result of the "Debtors" section. However, the first entry on the summary page relates not to the subtotal of the analytical page, but to the particular balance in the second account on page 86 R.

This allows us to generalize that our medieval colleague preferred the procedure of preparing the Trial balance in two stages: at first he collected the balances of unclosed accounts (primary indicators) at intermediate analytical pages; at the second stage, the subtotals were calculated for each analytical page and were transferred to the summary page (p. 92 L). After that, a grand total of all accounts of the "Debtors" section was calculated.

It is important to note that pages 87 R and 91 R are of a special character. The first one collected costs incurred, which had not been written off to expenses yet that were accounted in the General Ledger. Some separate records in this account do not let us identify the nature of costs, and most of them incline the authors that those were the costs that were going to be written off under to the "Household expenses". As for the page 91 R, there were the costs selected from the "Red Book of Expenditures" accounted on it. Further, it seems reasonable to seek for an answer to the question: Were the balances of unclosed accounts on the remaining intermediate pages collected in the order they were located in the General Ledger or did there exist any concrete linkages between the entries?

The total of the summary page of the "Debtors" section equalled to Lb. 2745 s. 6 d. 1 *to*.

The assignment of the page 92 R cannot be explained. The first entry says that the sum of Lb. 2745 s. 6 d. 1 *to* was transferred from the previous page, the second one – that the mentioned sum was transferred to the following page.

The page 93 L, from the authors' point of view, is the most important in the ledger. It is exactly because of it we can confirm that there is the summary of the balances of the creditors' accounts in the ledger on page 114. The page 93 L includes three records:

- Giovanni Farolfi and Co., **must have**, the 4 of August, we put where they **must have** here at c. 114, Sum Lb. 2762 s. 19 *to*;
- We **gave** to Giovanni Farolfi and Co. the 27 of June, year 300, we put where ser Giovanni Filippi and Co. **must give** here at c. 92, in sum. Lb. 2745 s. 6 d. 1 *to*;
- We **gave** to Giovanni and Co, the 4 of August, year 300, they **must give** for cost of this money until to this present day, for lb. 15 each 100, they sum. lb. 42 s. 6.

We have determined that in the Middle Ages the excess of the debit total of the Trial balance over the credit total was added to the reserve out of profit in most cases, while the reverse combination was deemed to be the accountant's receivables. Such a situation can be found in the example of the Farolfi company. The accountant was seeking for the unaccounted revenues and as a result of it there appeared the third entry.

4   Conclusion

In the course of the implemented research, thanks to the use of the cross-reference method, all the preserved pages of the ledger were identified and their virtual reconstruction was performed in a sequence that was adopted seven centuries ago. It allowed us to simulate the accounting system of "Debtors" section preparation in the Trial Balance in 1300.

The analysis of the accounting system and the procedures for preparing the Trial balance let us to verify its application in two stages. At the first stage analytical pages were formed on which the balances of the unclosed accounts of the General Ledger and other registers were collected. The result was generated on each intermediate page and was transferred to the summary table (c. 92 L). One indicator was transmitted to the summary table directly from the account (the second account on the carta 86 R) on the summary page.

The summary was transferred to the comparison page (carta 93 L), on which it was matched with another summary (credit) total posted on the missing carta 114.

The study of the minor preserved pages of the book will allow us to study in depth the accounts, the structure of the ledger, the accounting system and accounting procedure, which will increase our awareness of the history of the formation of double-entry bookkeeping.

References

Castellani, A.: Nuovi testi fiorentini del Dugento e dei primi del Trecento. T. 2. The ledger of Giovanni Farolfi & Company, 1299–1300, is transcribed, almost entire, Firenze (1952)

De Roover, R.: The development of accounting prior to Luca Pacioli according to the account-books of Medieval merchants. In: Littleton, A.C., Yamey, B.S. (eds.) Studies in the History of Accounting. Sweet and Maxwell, London (1956)

Lee, G.A.: The coming of age of double entry: the Giovanni Farolfi Ledger of 1299–1300. Account. Historians J. **4**(2), 79–95 (1977)

Martinelli, A.: The Origination and evolution of double entry bookkeeping to 1440. ProQuest Dissertations & Theses Global pg. n/a (1974)

Melis, F.: Storia della ragioneria. Zuffi, Bologna (1950)

Melis, F.: Osservazioni preparatorie al bilancio nei conti della Compagnia Farolfi, nel 1300. In: Studi di Ragioneria e Tecnica economica: Scritti in onore del Prof. Alberto Ceccherelli. Pubblicazioni dell'Università degli Studi di Firenze, Firenze, pp. 347–356 (1960)

Melis, F.: Documenti per la storia economica dei secoli XIII-XVI. Firenze (1972)

Smith, F.: The influence of Amatino Manucci and Luca Pacioli. BSHM Bull. **23**, 143–156 (2008)

The Early Cash Account Books

Dmitriy Aleinikov (iD), Mikhail Kuter$^{(\boxtimes)}$ (iD), and Artem Musaelyan (iD)

Kuban State University, 149 Stavropolskaya Street, 350040 Krasnodar, Russia
`prof.kuter@mail.ru`

Abstract. The operational accounting for cash transactions using double-entry method can be recognised as one of the formal indicators of double-entry bookkeeping. Unfortunately, the medieval cash registers were preserved in a limited quantity. In this matter, the exception is the company of Francesco Datini. The paper regards the practice of applying the earliest "Entrata e Uscita" in the company in Avignon. The method of calculation of the accounting cash balance is considered. Special attention is drawn to the fact that cash is entered daily or over several days, and the dispense cash transactions are implemented in specific amounts for each economic event.

Keywords: Datini company in Avignon · Cash · Entrata e Uscita

1 Introduction

Many years of practice in the archives of the medieval account books in Italy sometimes causes misunderstandings in terms of the feasibility of such studies nowadays. This is justified: the authoritative scientists carried out all the necessary research in the last century. In this case the names of Fabio Besta, Alberto Ceccherelli, Tommaso Zerbi, Federigo Melis, Raymond de Roover and others can be mentioned. However, the results of the latest research of Russian scientists turn out to be a vivid example for refutation of the stated opinion. In recent years, the significant results have been obtained: the detailed study of the medieval practice of impairment and depreciation charges (Gurskaya et al. 2016; Kuter et al. 2016), the early Profits and Losses accounts, the Trial balances, the use of balance account statements as reports sent from branches to the headquarters of the company (Kuter et al. 2017), the early experience of the formation and use of reserves out profit, and the early examples of adjusting entries were discovered. This paper describes the accounting practice of one of the earliest "Entrata e Uscita" in the company of Francesco Datini in Avignon (1367). The relevance of this issue is that the operational accounting of cash using double-entry method is considered to be one of the formal features of double-entry bookkeeping.

2 Review of Prior Literature

There is an opinion that the first researchers into the archives of Francesco Datini were F. Besta and A. Chekkerelli (Antonelli and Sargiacomo 2015, pp. 121–144; Sargiacomo et al. 2012, pp. 249–267). However, this is far from the case. Besta expressed his gratitude for the example prepared for him by his student Gaetano Corsani (Besta 1909, p. 318), who really worked with the archives (Corsani 1922, pp. 71–72). Chekkerelli, who mentioned the archives of Datini twice in his papers, also referred to the research of Corsani. The founder of the Tuscan view to the origin of the double-entry bookkeeping wrote: "G. Corsani in his study on the Datini's books on the same page 142, which is devoted to the company in Avignon, formed by Francesco di Marco, Toro di Berto, Boninsegna di Matteo, Tieri di Benci and Andrea di Bartolommeo (1367–1410)..." (Chekkerelli 1939, p. 31). This allows us to state that Chekkerelli did not work directly with the archival materials, but limited himself to the work of G. Corsani.

Marino Benelli and Cesare Guasti were the first to make an inventory of the preserved archival documents as they were studying those materials. Alas, the publications on their work were lost (Melis 1954, pp. 61–62).

Livi (1910), Dell Lungo (1897), and Nicastro (1914) are known as the first authors of the papers based on the archival research. Although, their works were descriptive in nature and had no generalisations or research conclusions.

The study of Bensa (1928) details what the account books of in Datini's companies were intended for. The researcher described "the results of many years research into the life of Francesco Datini, described the companies, society contracts, promissory note and bank contracts, transportation contracts, insurance and other commercial contracts" (Melis 1954, pp. 61–62) in his work. In addition to this, E. Bensa made a scientific contribution, which was expressed in his attempt to prove "the uselessness of research according to the territorial priority of the origin and development of double-entry bookkeeping". As well, he made an assumption that the double-entry bookkeeping simultaneously occurred and was developing concurrently in several trading towns (Bensa 1928, pp. 275–276) long before the same idea was given by de Roover.

After that E. Bensa sets before the researchers a more important task: "More logical than the study were earlier signs and more advanced systems in Venice than in Florence or Genoa, it would be the study, which could deduce the true origin and the priority between merchants' and bankers' records, who were undoubtedly the first to introduce the more accurate form of the books, which was able to guarantee the accuracy of the content" (Bensa 1928, p. 276). Lots of information on this issue is provided in the publication by Sangster (2016).

In 1974 Alvaro Martinelli wrote about the companies of Francesco: "he case of Francesco Datini from Prato is remarkable. He established a branch-office in Avignon, where from 1366 until 1401 he kept his books according to the old form, that is he used accounts with mingled sections; he introduced in his ledgers the new "Venetian method" only in 1401. This new form, which used accounts with lateral sections, had already been introduced into the accounting systems of the branch-offices of Genoa, Florence, Pisa, Barcelona, Majorca and Valencia which he established and managed with other partners from 1383 to 1393" (Martinelli 1974, pp. 191–192). At this point Martinelli makes reference to Melis (1962), thereby showing that he did not personally work with the archives of Datini.

Considering the cash accounting, Martinelli mentioned: "... book which is worthwhile mentioning here was the libra dell entrata dell' uscita or collections and payments book which followed the outline of similar ledgers used by public administrations. Its main characteristic was the division of collections, usually recorded in the first half of the book, and payments which were recorded in the second half. This created two simple or unilateral accounts which were joined only when cash had to be verified" (Martinelli 1974, p. 249). As well, there follows an allusion: "The earliest known example was found in the books of receipts and disbursements of the commune of Siena. This important set of books was begun in 1226 and records were kept, with a few interruptions, up to the XVII century".

With reference to Quaderno di Spese di Casa Segnato "B", Registro no. 421 (the Datini company in Pisa) Martinelli makes a generalisation: "Cash is another account with easily traceable cross-entries, which usually were recorded in a separate book, the libro "dell Entrata dell Uscita" or the "receipts and disbursements book", consequently it was deemed useless to indicate the position of its entries in the ledger. The lack of reference to the cash cross-entry was a main characteristic of the Tuscan accounting system until the end of the fourteenth century, which was found even in the ledgers of Francesco Datini in the first decade of the fifteenth century" (Martinelli 1974, p. 271).

Finally, attention should be draw to the fact that the cash flow was not recorded in the accounting system in the commune city of Genoa (1340). The only thing accounted was the cashier's accounts receivable. At the same time, several receivables accounts were opened for one cashier (Kuter et al. 2013).

3 Statement of Basic Materials

Raymond de Roover in 1956, with references to Corsani (1922, pp. 71–72) and Bensa (1928, pp. 409–413), described the bookkeeping in the Avignon company of Francesco Datini in the following way: "In the beginning of his career, while still residing in Avignon, Datini followed the prevailing Tuscan practice and kept his books in single entry, the ledger having split accounts, debits in front and credits in the rear, as in the del Bene and Peruzzi ledgers already mentioned. This form, which belongs to a transitional stage between the paragraph and the bilateral forms, is found in the libro giallo A, or ledger A bound in yellow [leather], of Datini's mercery business in Avignon (1367–1372). In an inscription on the front page, after the usual invocation to God, the Virgin, and the Saints, the book-keeper states explicitly that he intends to record systematically all entries involving a debit up to folio 150 and all those involving a credit from folio 151 onward to the end on folio 300. 17 Thus the debit of Niccolo di Bono's and Puccio Ricci's account is found on folio 2 verso and the credit on folio 151. To close the account, the total of the debit, or florins 270 14s. 8d. di camera, papal currency, was transferred from folio 151 to folio 2 and deducted from the total debit; the remaining balance was then collected in cash" (Roover 1956, pp. 140–141).

In the archive of Francesco Datini in Prato we had an opportunity to access the libro giallo A (the archival identification code – Prato, AS, D. № 2). Figure 1 gives an example, described by R. de Roover, in a scheme of the photocopies of real accounts. Figure 2 gives the diagram of the same fragment.

Tables 1 and 2 provide the translation of the entries of the mentioned accounts from Old Italian for a better understanding of the presented information.

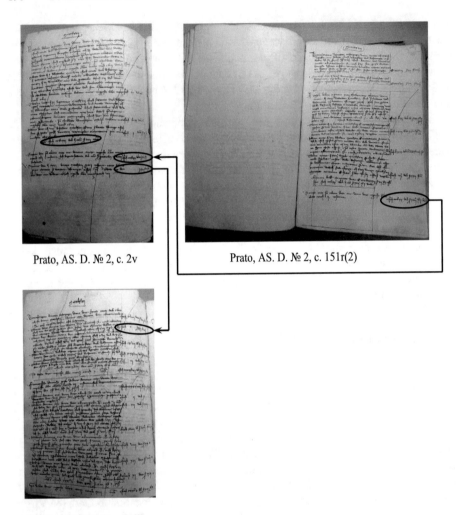

Prato, AS. D. № 2, c. 2v   Prato, AS. D. № 2, c. 151r(2)

Prato, AS. D. № 2, c. 7r(1)

Fig. 1. The Schematic representation of the process described by R. de Roover, in photocopies of real accounts

The translation of the accounts completely confirms the content of the article of R. de Roover. Indeed, the total, calculated on the debit entries (f. 271 of gold *di camera s. 8*) was decreased by the total on the credit entries of the account (f. 270 of gold s. 14 d. 10 *di camera*), which was transferred from Prato, AS. D. № 2, c. 151r(2).

It is worth focusing on the last recording in the account: They *gave* on March 21 1368 for us in Florence to Toro di Berto himself in cash s. 21 d. 5 *affiorini*, as the aforementioned Toro *must give* as in this book at c. 7, sum – s. 17 d. 8 *di camera*.

The cross-reference for the account of Toro di Berto (Prato, AS. D. № 2, c. 7r) informs that Nicolò di Bono and Puccio Ricci who live in Florence gave us March 31 1368 s. 21 d. 5 *affiorini* through Toro di Berto. The entry text says: "And he *must give*

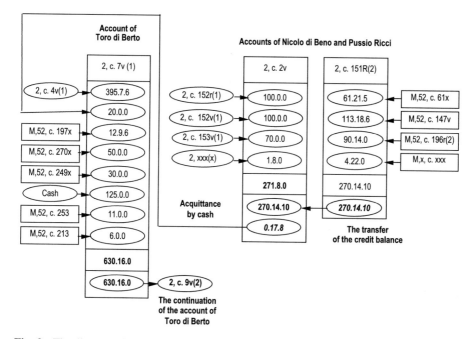

Fig. 2. The diagram of a part of the accounting process on the Debtors and Creditors accounts described by de Roover (1956, pp. 140–141)

on March 31 1368 s. 21 d. 5 *affiorini* he got for us in Florence from Michele di Bono and Piero Ricci as they **gave** in this book at c. 2". This allowed closing the account Nicolò di Bono and Piero Ricci (Prato, AS. D. № 2, c. 2v).

It is necessary to pay attention to a very important point: several currencies were used simultaneously in the calculations. So, the difficulty of the accountant's work consisted in knowing the rates of all these currencies and the ability to carry out the recalculations. Moreover, it was difficult to link records between the three registers: Memorial, General Ledger and Entrata e Uscita.

The cash accounting in the early books "Entrata e Uscita". It may seem paradoxical, but in the Datini companies the accountant had earlier learned to make analytical constructions for revealing the financial result (1366), than the first register for recording money was created (1367).

To the delight of the researchers, the earliest book of accounting for the inflow and outflow of cash "Entrata e Uscita" has been preserved in the State Archive of Prato under the index Prato, AS, D. № 92. Figure 3 shows the historical frontispiece of the book, where the name "Quaderno Entrata e Uscita" is clearly visible in the upper part of the leather binding. The entry on the left in the centre reads: "Book of Avignon, / 1367/1369". In the centre the arms of the owners are can be seen.

Figure 4 presents a photocopy of the first page of the book "Entrata e Uscita", on which, as a rule, the accountant placed the "oath".

Table 1. Translation of the account Prato, AS. D. № 2, c. 2v from the Old Italian language

1367	
Nicolò di Bono e Puccio Ricci **must give** on November 16 1367 f. 100 of gold *di camera* which they got in Florence from Antonio Guiardi and Giovanni d'Arrigo and Co. on October 11 of the same year, they made us pay that sum here in Avignon to Bottino and Bartolomeo Covoni f. 102 of gold s. 10 *a oro* s. 10 on November 11 for d. 2 ½%, the disadvantage/loss is up to us. There for, Bettino and Bartolomeo **must have** in this book at c. 152 on this day	f. 100 *d'oro di camera*
And they **must give** on December 10 1367 f. 100 of gold of Florence which they got in Florence from messer Nicolaio and Benedetto degli Alberti and Co. on November 6 of this same year in exchange of f. 102 of gold they made us pay here in Avignon to Laberto Lanberteschi and Co. on December 6 1367 and f. 2 of gold are for our disadvantage/loss. Therefore Lanberto and Co **must have** in this book at c. 152	f. 100 *d'oro di camera*
And they **must give** on January 11 1367 f. 70 of gold of Florence, they got in Florence from Davanzato and Manetto Davanzati on December 11 of that year in exchange of f. 72 of gold *di camera* which through their bill of exchange they made us pay on the same day in Avignon to Boninsegna of Matteo and Co., f. 2 are our disadvantage/loss for the exchange. Therefore Boninsegna and Co. **must have** in this book at c. 153	f. 70 *d'oro di camera*
And they **must give** on February 7 1367 lb. 1 s. 18 d. 5 *affiorini* they are for 4 vermillion sheepskings they charged us more, they are worth	f. 1 *d'oro di camera*
The sum is f. 271 of gold *di camera* s. 8	
They **gave** as they had or should have had as in this book at c. 151 f. 270 of gold *of camera* s. 14 d. 10 *di camera*	f. 270 s. 14 d. 10 *di camera*
They **gave** on March 31 1368 for us in Florence to Toro di Berto himself in cash s. 21 d. 5 *affiorini*, as the aforementioned Toro **must give** as in this book at c. 7	s. 17 d. 8 *di camera*

The book was opened on October 25, 1367 and consists of two separate parts:

- the first part (Entrata) was designed to account for cash inflows and took pages 2r–17r. Pages 17v–50v constitute the reserve left by the accountant for the first section;
- the second part (Uscita) was used to account for cash outflow and took pages 51r–93r.

The main source of information was the Memorial. Furthermore, the Memorial, applied in the company in 1367–1368, was the first register of such a type in the Francesco Datini Empire.

The accounting for the cash inflow was recorded not for each fact of the money income, but according to certain dates, which is clearly visible on the first information page of Prato, AS, D. № 92, c. 2r, a photocopy of which is shown in Fig. 5.

Table 2. The translation of the Prato, AS. D. № 2, c. 151r(2) into the English language

1367	
Nicolò di Bono and Puccio Ricci who live in Florence *must have* on October 21 1367 f. 61 of gold *di camera* and s. 21 d. 5 of Provence they are for one bale of iron gloves and small hammers (*martellini*) and pincers and vermillion sheepskins they send us from Florence until September 29 of the aforementioned year, and once all that was loaded on the galley it amounted to lb. 110 s. 8 d. 10 *piccioli* of s. 68 for florin, which are worth lb. 89 s. 15 d. 10 *affiorino*. They *must have* as in the Memorial A at c. 61, they are worth s. [—] for florin *di camera*	f. 61 of gold *di camera* and s. 21 d. 5 of Provençe
And they *must have* on January 3 1367 f. 165 *affiorini* they are for a shipment of 3 bales of Florentine merchandise they sent us until October 25 of the aforementioned year, we took from the Memorial Book A at c. 147, they are worth	f. 113 of gold *of camera* s. 9 d. 6 *camerali*
And they *must have* on February 8 1367 f. 130 s. 9 d. 6 *affiorini* for a shipment of one bale of Florentine merchandise they sent us until December 8 of the aforementioned year and so much it amounted the full cost with expenses, and then left Talamone as it can be seen in Memorial A at c. 196	f. 90 *di camera* s. 14 *piccioli*
And they *must have* on April 14 1368 for expenses made for the aforementioned bale from Florence to Talamone and expenses in Talamone as they charged uf through a letter from them we got the same day (it was written in Florence on March 18 1367, they amounted to lb. 7 s. 2 d. 6 *affiorini* they are worth)	f. 4 of gold s. 21 *of camera*
The sum is lb. 392 s. 7 d. 10 *affiorini* They are f. 270 of gold *di camera* s. 14 d. 10 *di camera*	
They had as they *gave* or they *should give* as in this book at c. 2	f. 270 of gold s. 14 d. 10 *di camera*

The totals were periodically summed up in each section and the remaining cash balance was calculated. Figure 6 presents a photocopy of the Prato, AS, D page. № 92, c. 2v, showing an example of the cash balance withdrawal in the book "Entrata e Uscita" on December 1, 1367, that is, a month and five days after the register was opened.

First of all, we would like to provide the translation of the top six lines on this page in Table 3. It draws attention to the fact that records on cash transactions were not made directly at the time of the cash inflow, but by the results of registration of the economic events in the Memorial. The "Entrata" comprised not the sums of individual transactions, but the accumulated totals by day, additionally in some cases over several days. In the cash register there was no reference to the page number in the Memorial or

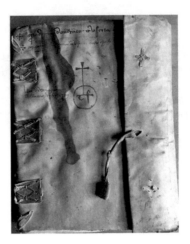

Fig. 3. The frontispiece of "Entrata e Uscita" of Francesco Datini's company in Avignon (1367), Prato, AS, D. № 92

Fig. 4. The first page of "Entrata e Uscita" (1367) (Prato, AS, D. № 92, c. 1r)

the General Ledger, which complicated the reconciliation with the registers of the business facts.

It is fairly obvious that the date served the only requisite-attribute, which linked "Entrata", and the Memorial (Prato, AS, D. No. 52). Thus, the entry in "Entrata" of November 22, 1367 can be verified only by adding the sums of the records up on the pages of the Memorial Prato, AS, D. № 52, pp. 90v–91r.

The link between the Memorial and the "Entrata" is more clearly visible through the single entry in the Memorial on November 30. Figure 7 amply shows the sum of Lb. 228 s. 12 d. 6, which is identical to the sum of the cash in the "Entrata" on Prato,

Fig. 5. The first page containing information in "Entrata e Uscita" (1367) (Prato, AS, D. № 92, c. 2r)

Table 3. The translation of the page Prato, AS, D. № 92, c. 2v of "Entrata e Uscita"

1367	
And on November 19 from the daily cash desk	Lb. 79 s. 8
And on November 20 from the daily cash desk	Lb. 17 s. 10 d. 11
And on November 22 from the cash desk	Lb. 32 s. 19 d. 8
And on November 23 from the cash desk	Lb. 13 s. 17 d. 7
And from on November 24 till 30 from the cash desk Lb. 77 s. 7	Lb. 77 s. 7
And on November 30 from the cash desk	Lb. 228 s. 12 d. 6

AS, D. № 92, c. 2V. In Figs. 6 and 7, the equal sums in the ellipses are marked with the dotted lines.

Figure 6 shows the way the total of the "Uscita" was transferred to the "Entrata" side and was written under the total of the "Entrata". Further, the remaining cash balance was calculated.

Table 4 generates an example based on the data of the first output of the cash balance in the book.

The first indicator stands for the total of the cash inflow calculated in the "Entrata" from October 25 to November 30, 1367.

The second indicator is the total of the cash outflow over the same period on the page Prato, AS, D. № 92, c. 54R in the "Uscita". The corresponding entry in the "Uscita" reads: "The sum of the records coming from October 25, 1367, as can be seen earlier in this book on six pages, is Lb. 923 s. 14 d. 3 in the currency of Provence, from the calculation s. 24 for each florine di grale. We place at the end of the Entrata, calculated over the same period of time as in this book".

The third indicator determines the cash balance.

At this point it is necessary to focus on the last entry on page Prato, AS, D. № 92, c. 2v: "This amount is put ahead on December 1st 1367". As can be seen in Fig. 6,

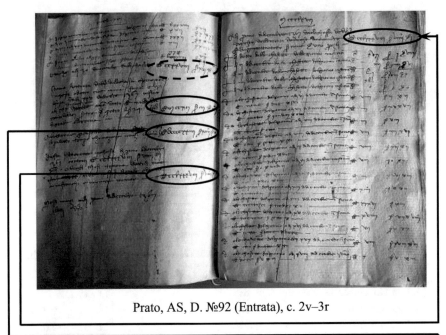

Prato, AS, D. №92 (Entrata), c. 2v–3r

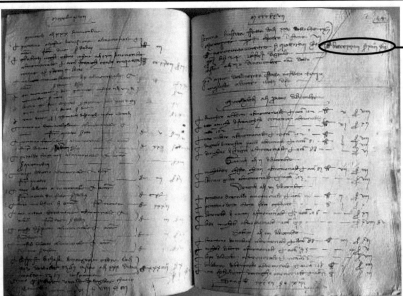

Prato, AS, D. №92 (Uscita), c. 53v—54r

Fig. 6. The calculation of the first intermediate cash balance in the book "Entrata e Uscita" (1367)

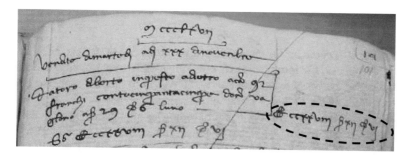

Fig. 7. The fragment of the page from the Memorial Prato, AS, D. № 52, c. 101r (1), indicating the amount of cash accrued in Entrata e Uscita on page Prato, AS, D. № 92, c. 2v

Table 4. The translation of the last three entries on Prato, AS, D. № 92, c. 2V in the book "Entrata e Uscita"

1367	
The sum of the cash inflow as it can be seen in this book until now, that is one page and what here above from October 25 1367 until November 30, is lb. 1212 s. 3 d. 4 at s. 24 for florin *di Grale o de Reina*	Lb. 1212 s. 3 d. 4
And we reduce for the cash outflow during the aforementioned time span as it can be seen further ahead	Lb. 923 s. 14 d. 3
It remains in cash on December 1st 1367 lb. 288 s. 9 d. 1 of Provence, they are in many currencies, that is in florins *di grali* and *di Reina* and Francs and in small change	Lb. 288 s. 9 d. 1
This amount is put ahead at December 1st 1367	

a new accumulation of the cash inflow starts with this amount. As for the cash outflow, its accounting starts from zero.

According to the full study of the "Entrata e Uscita" (Prato, AS, D. No. 92), the determination of the cash balance and the subsequent reconciliation with the cash inventory data of the cash desk were conducted monthly, on the last day of the month.

Another interesting aspect is that the records of the cash inflow over the month took one and a half pages, while the expense records were placed on six and a half pages. There is an explanation to it as the recordings in the "Entrata" were kept by the date (and sometimes over a few days), and in the "Uscita" – by each dispense cash transaction.

Finally, returning to the example described by de Roover (1956, pp. 140–141), on page Prato, AS, D. № 2, c. 7r it is said that Toro di Berto received 1368 s. 21 d. 5 affiorini from Nicolò di Bono and Piero Ricci on March 31, which is equivalent to s. 17 d. 8 di camera. We have made an unsuccessful attempt to find one of these amounts in the "Entrata e Uscita" on this date. The fact is that Prato, AS, D. № 2, c. 7r is the debit side of the Toro di Berto account, when its credit is Prato, AS, D. № 2, c. 156v (2). Accordingly, Toro di Berto did not bring cash to the cashier on a daily basis, but had to pay in the difference between the debit and credit components a year after.

4 Conclusion

The operational accounting for cash transactions by means of double-entry method is recognised as one of the formal attributes of double-entry bookkeeping. Unfortunately though, the medieval cash registers were preserved in a limited amount. The exception is the ones that belong to the company of Francesco Datini.

There existed three possible options for keeping accounting registers: in Entrata and Uscita: separate freestanding books; in one book; along with the book of merchandise.

The paper analyses the practice of use of the earliest Entrata e Uscita in the Datini company in Avignon. The method for the determination of the cash balance suggested the transfer of the total of the cash outflow from Uscita to Entrata and the creation of an entry under the total of the cash inflow. The calculated cash balance was placed as the first record in the accounting register for the new period in Entrata.

Some specific attention should be paid to the fact that the cash inflow was reflected in the accounting by day or over several days, when the dispense cash transactions were recorded in certain sums for each business operation.

References

Primary Sources

Archivio di Stato di Prato

Archivio di Stato di Prato, Prato, AS, D., Fondaco di Avignone, Company Toro di Berto, Francesco di Marco Datini e Tuccio Lambertucci nell'arte delle merci, Libri Grandi, 1367 −1373, No. 2

Archivio di Stato di Prato, Prato, AS, D., Fondaco di Avignone, Company Toro di Berto, Francesco di Marco Datini e Tuccio Lambertucci nell'arte delle merci, Memoriali, 1367 −1368, No. 52

Archivio di Stato di Prato, Prato, AS, D., Fondaco di Avignone, Company Toro di Berto, Francesco di Marco Datini e Tuccio Lambertucci nell'arte delle merci, Entrata e Uscita di Cassa, 1367−1370, No. 92

Secondary Sources

Antonelli, V., Sargiacomo, M.: Alberto Ceccherelli (1885–1958): pioneer in the history of accounting practice and leader in international dissemination. Acc. History Rev. 25(2), 121–144 (2015)

Bensa, E.: Francesco di Marco da Prato (Milan) (1928)

Besta, F.: La Ragioneria, 2nd edn. Rirea, Rome (1909). (In 3 Volumes), Facsimile Reprint (2007)

Ceccherelli, A.: l linguaggio dei bilanci. Formazione e interpretazione dei bilanci commerciali. Firenze, Le Monnier (1939)

Corsani, G.: I fondaci e i banchi di un mercante pratese del Trecento. Contributo alla storia della ragioneria e del commercio. Da lettere e documenti inediti, La Tipografica, Prato (1922)

De Roover, R.: The development of accounting prior to Luca Pacioli according to the account-books of Medieval merchants/R. de. Roover. Littleton, A.C., Yamey, B.S.: Studies in the History of Accounting, London, pp. 114–174 (1956)

Del Lungo, I.: Francesco di Marco Datini mercante e benefattore. Giaclietti, Prato (1897)

Gurskaya, M., Kuter, M., Papakhcian, A., Musaelyan, M.: Specific features of depreciation accounting at the end of the 12th – early 13th centuries. In: 5th International Conference on Accounting, Auditing, and Taxation (ICAAT 2016). https://doi.org/10.2991/icaat-16.2016.10

Kuter, M., Gurskaya, M., Bagdasarian, R., Andreenkova, A.: Depreciation accounting in Francesco Datini's Companies. In: 5th International Conference on Accounting, Auditing, and Taxation (ICAAT 2016) (2016). https://doi.org/10.2991/icaat-16.2016.26

Kuter, M., Gurskaya, M., Andreenkova, A., Bagdasaryan, R.: The early practices of financial statements formation in Medieval Italy. Account. Historians J. 44(2), 17–25 (2017)

Kuter, M.I., Gurskaya, M.M., Sidiropulo, O.A.: The Genoese Commune Massari's Ledger of 1340: the first computer modeling experience and its results. J. Modern Account. Audit. 9(2), 212–229 (2013). ISSN 1548–6583

Livi, G.: Dall'Archivio di Francesco Datini mercante pratese. F. Lumachi, Firenze (1910)

Martinelli, A.: The origination and evolution of double-entry of bookkeeping to 1440. ProQuest Dissertations and Theses; 1974; ProQuest Dissertations & Theses Global pg. n/a.н (1974)

Melis, F.: L'archivio di un mercante e banchiere trecentesco: Francesco di Marco Datini da Prato. Moneta e Credito, VII (1954)

Melis, F.: Aspetti della vita economica medievale (studi nell'archivio Datini di Prato). Monte dei Paschi di Siena, Siena (1962)

Nicastro, S.: L'Archivio di Francesco di Francesco di Marco Datini in Prato, Rocca S. Casciano, L. Cappelli (1914)

Sargiacomo, M., Servalli, S., Andrei, P.: Fabio Besta: accounting thinker and accounting history pioneer. Account. Hist. Rev. 22(3), 249–267 (2012)

Sangster, A.: The genesis of double entry bookkeeping. Account. Rev. 91(1), 299–315 (2016)

The Impact of Financial Crisis on the Role of Public Sector Accountants

Tatiana Antipova[✉]

Federal State Budgetary Educational Institution of Higher Education,
Perm State Agro-Technological University Named After
Academician D.N. Pryanishnikov, Perm, Russia
antipovatatianav@gmail.com

Abstract. The main idea of current research is that role of Accountants may change under the influence of financial crisis. Change is considered in three aspects: timing of the accounting (1), the role of accountants (2) and accounting standards (3). One method to determine the impact of financial crisis is to carry out surveys. This paper indicates the responses to the survey of government and public sector accountants in three regions of the Russian Federation. The responses to Russian survey compared with the responses to international survey. The survey was carried on the same issues that were raised during questioning conducted by CIMA among commercial finance professionals in manufacturing, financial services, and wholesale and retail trades. Author's study found that in public sector governance (1) the impact of the crisis on the time reported is insignificant, (2) average 4% of respondents determines crisis as a reason for changing the role of accountants, and (3) due to the need to reduce fiscal restructuring and the polarization of the budget network, intensified standardization state records in the Russian Federation.

Keywords: Financial crisis · Timing of the accounting · Role of accountants
Accounting standards · Accounting trends

1 Introduction

The crisis has shown that the global economy needs effective urgent support. Moreover, the crisis can be overcome only identifying and addressing its real causes. It is imperative to understand the causes of the crisis in order to avoid equally significant crises in the future. Crises arise as a consequence of an unconscious side of society. Unconscious in the sense of ignorance or forget fullness main cause of crises and not taking action. The cause of the crisis may not be the absence or elimination of something, because what is not is not working and could not leave some sort of impact. The cause of the crisis is the fact that there is something that has great power and mighty works, because in order to mismatch of millions of people need a very large force. Man, this force does not afford and does not depend on the will and desires of people. Moreover, it opposed them. However, the cause of the crisis is not a natural but a social phenomenon, which is generated by the activities of society.

© Springer International Publishing AG, part of Springer Nature 2018
T. Antipova and Á. Rocha (Eds.): MosITS 2017, AISC 724, pp. 208–214, 2018.
https://doi.org/10.1007/978-3-319-74980-8_19

Science more than a hundred years ago was proved that the broken or inconsistent linkages between producers and consumers are the main cause of the crisis. Disengagement from active management of power led to the decline of a number of essential management functions such as planning, coordination, accounting and control. Accounting for cash flows has been separated from the material flow and address less. Material flows can have one address, or even might not be, but the money for them - their addresses.

This spoil account spawned an explosion of economic crime: the appropriation of state property, interception of money on false advice, withholding tax, unjustified transfer money abroad, cashing, money laundering and peculation, etc. The idea of this account is, in particular, to take into account simultaneously and material and financial flows. Accounting for material and cash flows should be combined into one system. There should also be clear and unified ideology that questions about what connection, as they consider and how to automate their accounting, understood equally by all the services of power structures.

Under the influence of financial crisis, accounting trends may change. It poses the question of whether there are universal role models that everyone can accept. Singleton-Green points out "conventional theory and practice is ill-suited to the challenges of the modern environment, and that accounting practices and corporate behavior are inextricably connected with many allocative, distributive, social, and ecological problems of our era. From such concerns, a new literature is emerging that seeks to reformulate corporate, social, and political activity, and the theoretical and practical means by which we apprehend and affect that activity. The major decisions on accounting and its regulation are unavoidably political because they are government policy issues that affect the general welfare as well as the welfare of particular individuals and groups" (Singleton-Green 2010).

However, in their paper Van der Stede and Malone (2010) argue that accountants kept the books and assured that statutory requirements were met. Even as the role became more complex with innovative organisational models, a proliferation of regulations, and a glut of new financial instruments, the focus on the books remained. And those trends, according to which many professional accountants can manage risk at a high level of strategic management, have not yet found distribution in the area of budget accounting in Russia. It shows results of the survey are listed below. The idea of questioning was born as a result of the study materials provided by CIMA at the World Congress of Accountants in Kuala Lumpur in November 2010 (Van der Stede and Malone 2010). CIMA has conducted an online survey in manufacturing, financial services, and wholesale and retail trade. They analyzed the results of research on the Anglo-Saxon markets (Australia, New Zealand, North America, United Kingdom) and Asia (China and India).

2 Data and Methodology

One method to determine the impact of financial crisis is to carry out surveys. This paper indicates the responses to the survey of government and public sector accountants in three regions of the Russian Federation. The responses to Russian survey compared with the responses to international survey. The survey was carried on the same issues that were raised during questioning conducted by CIMA among commercial finance professionals in manufacturing, financial services, and wholesale and retail trades.

For current study survey covered 292 respondents, who were in training courses in the Russian Federation. Here are the results of the processing of questionnaires, filled with 71 respondents from Moscow, 188 respondents from Sankt Petersburg and 33 respondents from the city of Arkhangelsk. In Arkhangelsk government establishments' accountants were interviewed during December 2010, in Moscow and Sankt Petersburg – during December 2010 and May 2011.

Questionnaire is based upon the results of the survey conducted by Van der Stede and Malone (2010) in order to compare our results with the global trends. The questionnaire consists of questions on ranking the time allocation by accountants for the following activities: «**accounting operations**; **external reporting**; **management accounting**; **management support**; **management information systems**; **other**» (Van der Stede and Malone 2010, p. 2).

The questions were slightly changed in accordance with Russian conditions. But the bulk of the issues have not changed, and time reported by type of activity was divided into:

Accounting operations (processing of source documents, data loading to the computer, accounts receivables and payables, internal financial reports);

External reporting (financial, fiscal and statistical reports, in particular, the interaction with the treasury or the credit institutions, financial risk management, including internal audit, compliance with regulations);

Management accounting (preparation and interpretation of management accounting information, such as forecasting, budgeting, costing and cash flow management);

Management support (identifying and analyzing strategic options, decision support, planning and tracking of key performance indicators, testing, strategic management accounting and business management);

Management information systems (development, implementation and maintenance of management information systems);

Other (personnel management, training and other activities).

Responses to our survey indicate that Russian governmental accountants are dividing their timing of the accounting a variety of responsibilities. Each type of work respondents scaled from 1 to 5 that correspond to the importance of work identified in the questionnaire. As a result, the average score obtained from respondents has been calculated.

And also we have been interested in responses concerning the reasons of changing roles of accountants in recent years. To assess the changing role of accountants the following question has been asked in the questionnaire: «Reasons for the changing role of accountants in recent years: Legal requirements; Professional and ethical requirements of the self-regulating non-profit organizations; Requirements from employers; Financial crisis; Other».

3   Study Result

Russia Results of the study examined three aspects: *timing of the accounting (1), the role of accountants (2) and accounting standards (3)*.

In order to determine *timing of the accounting (1)*, we asked the accountants how employee hours were spent within the finance function of their organisations. Russian results were obtained by summing these three regions – Moscow, Sankt Petersburg and Arkhangelsk. From the overall respondent pool, accounting operations (traditional tasks and transaction processing) account for more than half the time spent by respondents on the job, 51%. It is interesting also to compare Russian results with CIMA survey (Van der Stede and Malone (2010)) of FTEs within the finance function (Exhibit 1). The comparison was performed only on the current (actual) distribution because the data on preferences are CIMA not in absolute terms as well as the difference between the required, preferred, and current distributions of the time (roles).

Figure 1 shows that Russian's respondents put more manpower generally against accounting operation and external reporting. The governmental accounting sector is the most conservative and inertial structure in comparison with commercial structures and less prone to all kinds of crisis. The results of our survey also underscore the need for accountants to remain true to the traditional duties of accounting operations. The fundamentals of accounting will remain vitally important and will underpin all other activities at every level of an establishments.

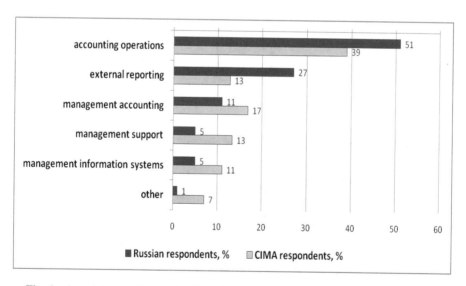

Fig. 1. Comparison of the results about time reported Russian and CIMA respondents.

CIMA's respondents placed more against management accounting, support and information system. Part of this balance is a result of the financial crisis. For example, according to CIMA accounting operations is a key priority in the allocation of working time to 39% of respondents, and in Russia – at 46%. It is possible that differences between the survey results are due to the different categories of respondents. Because CIMA processed results of the survey FTEs within the finance function in manufacturing, financial services, and wholesale and retail trade, and our survey involves only accountants from government agencies. However, for both commercial and governmental accounting the main function is to present reliable facts of economic activity. But this is the next research task – to expand the survey area and the number of its participants.

To assess the changing *role of accountants (2)* the question about the reasons for the changing role of accountants in recent time has been asked. The choice of answers was the following: Legal requirements, Professional and ethical requirements of the self-regulating non-profit organizations, Requirements from employers; Financial crisis and "Other". Answers to this question are presented in Fig. 2.

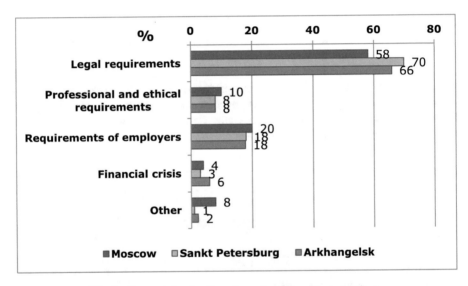

Fig. 2. Reasons for the changing role of Russian accountants.

Comparing the survey results in 3 Russian regions (Moscow, Sankt Petersburg and Arkhangelsk) on the assessment of the reasons for the changing role of accountants in recent years (% of preferences) is presented in Fig. 2.

Figure 2 shows that most respondents (58%, 70% and 66% respectively) mentioned as the reason of the changing role of accountants the legal requirements. But the legal requirements for an accountant have not been changed for the last 14 years. Federal Law "On Accounting" was adopted in 1996 and since then has not been drastically changed. Apparently, the respondents had in mind the requirements of the Regulations for the budget accounting. Because these instructions have been changed during last 5 years for 4 times. Thus, this is probably considered as changing legislative requirements. As such the requirements to the role of accountants have not been changed. The organization of accounting in government establishments have been reformed. And in 2011 it will be again modified.

Changes since 2011 due to restructuring of the budget network. Change of Russian governmental accounting standards (3) is presented in Table 1. The crisis, the Russian government should reduce budget expenditures. One of the measures to reduce budget expenditures is to reduce the number of government establishments, contained within the budget. To do this, government agencies are classified into three types: state-owned, budgetary and autonomous establishments. Entirely on the budget financing are only state-owned establishments (in 2010, was created 4943 state-owned establishments from 245272 the total number of government establishments). Autonomous and budgetary establishments transferred to the regime of subsidies. That is, from the budget, they will receive subsidies. Subsidies are to be based on performance.

Table 1. Changing of Russian governmental accounting standards.

2010 Government accounting standards				
Order of the ministry of finance RF by 30.12.2008 148n				
Currently government accounting standards				
Order of the ministry of finance RF of **01.12.2010 N 157n**	Order of the ministry of finance RF of **15.12.2010 N 173n**	Order of the ministry of finance RF of **06.12.2010 N 162n**	Order of the ministry of finance RF of **16.12.2010 N 174n**	Order of the ministry of finance RF of **23.12.2010 N183n**
The total for all government establishments	The total for all government establishments	For state-owned establishments	For budgetary establishments	For autonomous establishments

The Ministry of the Finance of the Russian Federation has always been in charge of defining rules of public sector accounting and reporting in Russia through issuing many reduction of "instructions" for how registration, measurement and reporting should be done. The Ministry of Finance has not changed the nature of regulation and still regulates all accounting procedures through detailed instructions. Due to the restructuring of the budget network, the Finance Ministry RF has approved for each new type of institutions his instructions on accounting. Therefore, in 2011 there was a sharp increase in the number of regulations governing accounting in government establishments. Instead of a single statement in 2010, in 2011 were approved almost identically five instructions (see Table 1).

4 Conclusions

The study result found that in public sector governance (1) the impact of the crisis on the time reported is insignificant, (2) average 4% of respondents determines crisis as a reason for changing the role of accountants, and (3) due to the need to reduce fiscal restructuring and the polarization of the budget network, intensified standardization state records in the Russian Federation. In the public sector impact of the crisis is not so obvious, because public sector institutions have functioned for budget funds, which are practically not diminished. For instance, 2008–2009 Russian federal budget expenditures up from €182690 billion to €200198 billion, respectively.

The results of our questionnaire show that the role of government accountants in Russia until now is not developing in the direction of strategic management of the institution. The governmental accounting sector is the most conservative and inertial structure in comparison with commercial structures and less prone to all kinds of crisis.

The results also emphasize that governmental accountants remain faithful to the traditional accounting responsibilities – accounting operations, external reporting, invoice processing and so on. Even when they want to improve their competence, even then the tradition of accounting will continue to be vital and will underpin all other activities at all levels of an organization.

But as global financial crisis has clearly shown, major corporations are in need for financial professionals who understand how to manage risk, cash flow, financial instruments and other complex functions and can offer a change of strategy and, indeed, to govern the organization on an equal footing with the heads. Those who continue to exclusively use old skills are likely to remain behind. And those who accept the expansion of their roles and responsibilities will be in demand.

References

Order of the Ministry of Finance of the Russian Federation of 30.12.2008, No. 148n On Approval of Instruction on Budgetary Accounting

Order of the Ministry of Finance of the Russian Federation of 01.12.2010, No. 157n On Approval of the Uniform Chart of Accounts for Government (Municipal) Establishments and Instructions for Its Use

Singleton-Green, B.: The communication gap: why doesn't accounting research make a greater contribution to debates on accounting policy? Acc. Eur. 7(2), 129–145 (2010)

Van der Stede W.A., Malone, R.: Accounting trends in a borderless world. CIMA. World Congress of Accountants (2010)

The Determination of Profit in Medieval Times

Alan Sangster[1] , Mikhail Kuter[2](✉) , Marina Gurskaya[2] ,
and Angelina Andreenkova[2]

[1] University of Sussex, Falmer, Brighton BN1 9RH, UK
[2] Kuban State University, 149 Stavropolskaya Street, 350040 Krasnodar, Russia
prof.kuter@mail.ru

Abstract. This paper presents the results of research into the early practices adopted to identify financial performance. Focusing on the Pisan account books of Francesco Datini from 1383–1386, it reveals the special features of the calculation of profit or loss on each transaction. This, in its turn, is consolidated into the calculation of the overall profit or loss. This consolidation step was performed by transferring these profits and losses into a *mercanzie* account, literally, a merchandise account. All other income and expenditure that did not form part of a venture or a consignment was also transferred to this account. When balanced, it revealed the net profit for the period, just as today we find that in the statement of profit and loss, or income statement. The sophistication of the approach demonstrates the high level of the accounting culture in this medieval business, and suggests that this early precursor of the modern profit and loss account may have emerged earlier than the concept of a Trial Balance, rather than simultaneously as has previously been thought.

Keywords: Financial performance · Determination of profit
"Profits and losses" account · Trial Balance · The Datini company

1 Introduction

This paper concerns the discovery of one of the earliest examples of a precursor to a modern Statement of Profit and Loss. In the last quarter of the 20th century, research into the archives of medieval account books took on a secondary priority. Many believed that anything worth knowing about this period of accounting history had already been investigated and described in detail. However, many years of research in the archives of Florence, Genoa, Venice and Prato, as well as in the Central Library of Florence has revealed to these researchers that this is far from the case. The authors' development of method of logical and analytical reconstruction (MLAR) facilitates the representation in block-diagram models of the entire accounting process as recorded in the account books, covering all steps from original entry to production of financial statements.

2 Prior Literature

Several scholars have written about the various enterprises of the 14th and 15th century international merchant, Francesco di Marco Datini (c.1335–1410) many whose business records and correspondence were preserved in his former home in Prato, a town

© Springer International Publishing AG, part of Springer Nature 2018
T. Antipova and Á. Rocha (Eds.): MosITS 2017, AISC 724, pp. 215–224, 2018.
https://doi.org/10.1007/978-3-319-74980-8_20

not far from Florence. Many believe that the first to do so was Fabio Besta, since he was the first who wrote about it in the historical accounting literature, (Besta 1909, pp. 317–320). Yet, it was one of Besta's former students, Gaetano Corsani, who was the first to examine the records of Datini's businesses. Corsani later used the results of his research when writing his own work (Corsani 1922, pp. 83–85). Another former student of Besta, Alberto Ceccherelli, wrote about the archive of Datini's business records in Prato (Ceccherelli 1913, 1914a, b, 1939). Others have written about the records held in this archive, including Bellini (Melis 1954), Carradori (1896), Guasti (1880), Livi (1910), Nicastro (1914) and Bensa (1923, 1925, 1928). Biographies Datini are of special interest for the researchers as they use as their sources large number of preserved personal and business letters and documents: (Lungo 1897; Origo 1957; Nigro 2010). In the publications of the past century we did not manage to find any examples, which would unearth the accounting practice including related to double-entry bookkeeping. Even while describing the preserved accounting registers of the Datini companies, such accounting historians as Martinelli (1974) and Zerbi (1952) did not consider the detailed mechanism of accounting. Only Federigo Melis and Raymond de Roover devoted any attention to the accounting procedures.

One of the explanations to it could be the monopolization of the State Archive of Prato by Federigo Melis in the mid-20th century. However, aside from describing the companies and linking them to some definite time periods, he identified little that might enrich the history of the development of accounting.

For example, Raymond de Roover, wrote (1956, p. 141):

> This system [of bookkeeping was] altered in a later ledger (1383–1386), [in a ledger] belonging to Datini's Pisan branch. In it the personal accounts for receivables and payables are in bilateral form, but merchandise expense, and profit-and-loss accounts continue to have the credit beneath instead of beside the debit. The presence of accounts for operating results has led to the conclusion that this ledger is in double entry, but Professor Zerbi, who has examined it carefully, challenges this opinion because of the absence of a cash account, though there is a cash-book which might have been used as a substitute. Combining the cash-book with the ledger would enable the book-keeper to strike a balance, provided, of course, that he observed the rules in other respects. The argument [concerning whether or not this was an example of double entry], therefore, is not decisive.
>
> Professor Zerbi also intimates that none of Datini's account-books contains a clear-cut example of double entry. On the contrary, after 1390 this system was certainly applied in most of the Datini branches abroad and at his main office in Florence.' Branch managers were expected to send regularly a copy of the balance sheet to headquarters, and several of such copies are still preserved in Prato. They show that the books were in balance save for small errors which the book-keepers often did not bother to trace but preferred to adjust by posting them to Profit and Loss.

The Datini enterprise in Pisa was among the first to move to bilateral accounts with opposing debits and credits. This is where the *mercanzie* account appeared for the first time extended to include the creation of reserves out of profit, making adjusting entries (Kuter et al. 2017), and recording impairment of non-current assets (Kuter et al. 2016). The Pisa company also prepared a Trial Balance within the ledger containing, within each entry, the folio references for the debits and credits in the closed and opened ledgers. Copies of the balance account were also prepared and sent to the geographically distant owners. None of this detail has previously been identified.

3   The Archival Material Used in This Study

The Ledger of the first Datini enterprise in Pisa (Campione Giallo "B" – the identity code of the archive of Prato, AS. D. № 357) is known to many researchers as a book with a yellow binding, and is often referred to as the 'yellow book'. It comprises of 476 pages, many accounts for obligations and settlements with counterparties (suppliers and customers), and others for the sale of goods and the accounts for the consignments and ventures for each of which the profit or loss is obtained.

The accounts in the Ledger differ according to their nature. The personal accounts of agents and correspondents were kept according to the bilateral format often described as the 'Venetian method'. In this format, the debit was placed on the left of two facing pages, and the credit on the right. In the case of goods and consignment accounts, an older format is used: mingled accounts (see, Martinelli 1974). Mingled format typically has the debit above the credit on the same page, rather than on opposite pages.

The accounts for sale of goods were used both in Venetian and Tuscan formats (the debit is in the upper part of the account, the credit – in the lower part, the operating result was written under the credit total despite its final outcome – profit or loss).

The accounts for the accumulation of operating results (one particular account was meant to accumulate operational profits and the other one – operating losses for the same type of activity, for example, for the sale of goods) were one-sided, i.e. there were only debit or credit entries placed. For this reason, the Tuscan account format was used. In order to accumulate operating profits, the credit of the account took a separate page on the spread. The debit of the account was written under the credit total and was intended to close the account and transfer the credit total to a new account to some free space in the back of the book. For the accounts for the accumulation of operating losses the sequence of entries in the account was the opposite.

However, already a year later there appeared a problem positioning of the books due to their purpose. Among the books intended for the systematisation of the accounting data are: The book of debtors and creditors or the General Ledger (Libro Debitori e Creditori or Libro Grande) – Prato, AS. D. № 358; the book of trade (Libro di Mercanzie – Prato, AS. D. № 377); the book of cash (Entrata e Uscita) – Prato, AS. D. № 403.

The first book – General Ledger Prato, AS, D, № 358 – contains the accounts for settlements with customers and suppliers. It was foliated from1R to 452R. Its entries terminate by 1385. The key feature of the records in the book is in the fact that all the debtors and creditors accounts were constituted in the Tuscan format.

The main book with the accounts for sale of goods is Prato, AS, D, № 377.

The Tuscan format was mainly used for the venture accounts (the accounts for sale of goods). The book includes the account for the accumulation of operating results as well. The most important peculiarity in this narration is that one of the earliest accounts for profits and losses prepared according to the data from 1383 to 1386 is placed on cartas AS. D. № 377, c. 73v–76r.

The Book of cash receipts and expenses ("Entrata e Uscita") is Prato, AS, D, № 403. Starting with carta 2r up to carta 87r there are the records of cash incomes; cartas

87v – 118r are blank pages (back-up pages for cash incomes accounting); cartas 87v – 201r contain the recordings of cash outflow. This book was kept in 1382 before the application of double-entry bookkeeping method. Nevertheless, it seamlessly integrated into the accounting system oriented towards double entry. Furthermore, it was being used after 1387, when at the end of 1386 after the three-year accounting cycle the rest of the books were closed.

In addition, there were books intended for pre-recordings. First of all, they include Memoriale. In the first enterprise (1383–1386) those were Memoriale "B" (Prato, AS. D. № 367) and Memoriale "C" (Prato, AS, № 368). However, the primary records were not only registered in the Memorial (as it is said in Luca Pacioli's treatise or in the first Venetian books), but also synthesized them for the General ledger. The total transaction amount was calculated on each purchase or sale transaction registered in Memoriale and, then, correspondence of accounts was performed.

The books Prato, AS, D, № 388 "Quadernoi di Spese di Mercanzie" and Prato, AS, D, № 387 "Quaderni di Ricevute e Mandate di Balle" are of special interest. They were designed for analytical accounting of the receipt of goods, the determination of the full value of the consignment and its transfer to the debit of the control account for the sale of the goods batches.

It is necessary to proceed to a well-established system of accounts designed to arrive at profit or loss. The use of MLAR has allowed the authors to trace the accounting system from each sale transaction to the overall profit or loss of 1383–1386. Additionally, there was made an emphasis on the intermediate results for each type of activity.

The figure shows the diagram of the determination of profit for the period of 1383–1386. The accounts in the scheme are arranged in the following order. The right series of accounts stand for the accounts for the sale of goods, in which the operating result for each sale was calculated (Fig. 1). As previously stated, such accounts were also placed in the General Ledger (Prato, AS, D, № 357) and in the book of trade (Libro di Mercanzie - Prato, AS, D. № 377). In the first case, both formats (Venetian and Tuscan) were used to compile a sales account, in the second case only Tuscan format was applied.

The Table 1 represents the translation of the first (upper) account on Prato, AS, D, № 377, c. 17r from Old Italian. The full cost of the sold consignment of the joint activity of the company Francesco Datini and Ambrogio di Bino (sum of f. 105 s .2 d. 9 of gold) is determined in the upper (debit) part of the account.

The sales revenue is reflected (f. 110 s. 7 d. 7 in gold) in the lower (credit) part of the account.

Further, the accountant distributes the operating profit from the transaction (f. 5 s. 4 d. 10) equally between the Datini company and the partner who participated in it: "The half is for Ambrogio, that is f. 2 s. 12 d. 5 we put at the Libro Giallo [Yellow Book] at c. 282, the remaining is f. 2 s. 12 d. 5, we put at the account Profit of merchandises in this book at c. 44".

The figure clearly shows the transfer of the profit share withdrawn by Datini (the upper entry on Prato, AS, D, № 377, c. 17r) for accumulation on Prato, AS, D, № 377, c. 44r. Concerning the profit share of the partner, it was carried forward to the personal account of Ambrogio di Bino in the General ledger on Prato, AS, D, № 357, c. 282v.

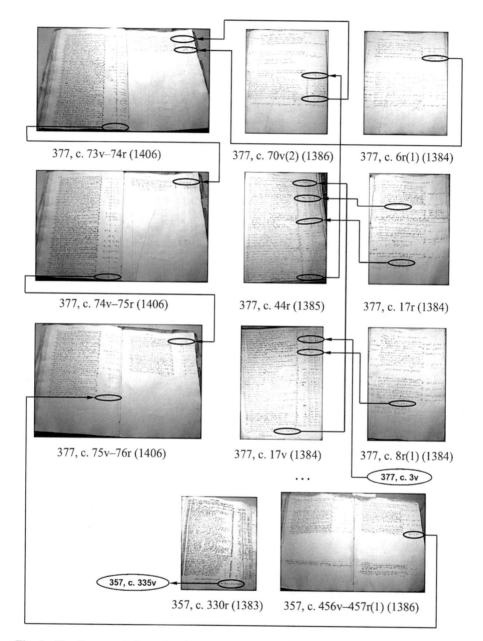

377, c. 73v–74r (1406)   377, c. 70v(2) (1386)   377, c. 6r(1) (1384)

377, c. 74v–75r (1406)   377, c. 44r (1385)   377, c. 17r (1384)

377, c. 75v–76r (1406)   377, c. 17v (1384)   377, c. 8r(1) (1384)

377, c. 3v

. . .

357, c. 335v

357, c. 330r (1383)   357, c. 456v–457r(1) (1386)

Fig. 1. The diagram of determination of profit in photocopies of real accounts (the individual enterprise of Francesco Di Marco Datini in Pisa, 1383–1386)

It is necessary to note that the sales accounts never took more than one spread in the General ledger or more than one page in Libro di Mercanzie. In the centre of the diagram there is a serial (dynamic) line of the accounts "Profit on merchandise". The

Table 1. The translation of the sales account Prato, AS, D, № 377, c. 17r.

	Purchase of some iron wares and [*aguti*] nails for Palermo for us and Ambrogio di Bino		
Lib. 1000	Of nails of many types in 4 bales for f. 4 ½ [each 100]	We bought these things from Piero	f. 97 s. 9 d. 9
Lib. 750	Of *gangheri* [iron wares] of many types in 1 bale for f. 6 s. 60 [each 100]		
Lib. 1830	Of *bomerali* [iron wares] and silver for f. 10 s. 3		
	From Del Tignoso, and the total amount is, as in the Memorial C c. 279		
F. 37 +	To weigh *aguti* and *gangheri* and load them on a wagon s. 11, for the bill […] they carried to the shipment s. 20, for transport from Pisa to Liborno f. 2 s. 10, for customs of Livorno s. 62, for 100 of ropes and other kind of cordage f. 1 s. 16, for insurance in Pisa Livorno for f. 7 at 3% f. 2 s. 63, for … of insurance s. 5 …	f. 7 s. 14 *a oro*	
	Sum f. 105 s. 2 d. 9 of gold		
F. 3 cantari t° 98	Of *aguti* that Ambrogio di Bino sold in Palermo as in the Memorial at c. 439, they were, net of expenses	f. 50 s. 14	
F. 1 cantari t° I	Of *gangeri* that said Ambrogio sold as at the Memorial at c. 439, they were, net of expenses	f. 19 s. 2 d. 7	
f. 100	*Bomerali* that said Ambrogio sold as at the Memorial at c. 439, they were, net of expenses	f. 40 s. 11	
The sum is f. 110 s. 7 d. 7 of gold			

The rest is f. 5 s. 4 d. 10 of gold. The half is for Ambrogio, that is f. 2 s. 12 d. 5 we put at the *Libro Giallo* [Yellow Book] at c. 282, the remaining is f. 2 s. 12 d. 5, we put at the account Profit of merchandises in this book at c. 44

diagram shows four out of eight accounts in the line. Regardless of the place of registration, all the accounts were executed in the Tuscan format. The line of accumulation of operating results begins in the General ledger, which was opened in 1383. Four accounts were alternately opened in it – Prato, AS, D, № 357: c. 330r (the transferred balance – f. 384 s.3); c. 335v (the transferred balance – f. 800 d. 9); c. 338r (the transferred balance – f. 1079 s 11 d. 11); c. 382v (the transferred balance – f. 1346 s. 7 d. 4). Every time when the following account was closed, the accumulated total was transferred to the newly opened account as its opening balance.

In the *Libro di Mercanzie*, that was opened a year after the official presentation of the company in Pisa (Prato, AS. D. № 377), there are four accounts that served as the continuation of the previous accounts: c. 3v (the transferred balance – f. 1667 s. 4 d. 6); c. 17v (the transferred balance – f. 2377 s. 8 d. 2); c. 44r (the transferred balance – f. 3142 s. 4 d. 7); c. 70v(2) (the transferred balance – f. 3215 s. 3 d. 11).

The accumulated total from the eight accounts (f. 3215 s. 3 d. 11) was transferred to the second position on the spread of the account for profits and losses (Prato, AS, D, № 377, c. 74r): "And they **must have** f. 3215 s. 3 d. 11 *a oro* for profit of many trades we put in this book at c. 76".

In the studied accounting system the operating results were transferred to the "Profits and Losses" account both as accumulated and individual indicators. The figure shows the account on folio Prato, AS, D, № 357, c. 456v–457r, which revealed the operating loss in the amount of f. 37 s. 15. The indicator was carried over to the 15th (last) position on the third folio (Prato, AS, D, № 377, c. 75v) of the "Profits and Losses" account: "And they must give f. 32 s. 15 a oro for letters sent and received from Genoa to Avignon and from Avignon to Genoa by the hands of Bruno di Francesco and Francesco di Marco himself paid for us in the Yellow Book at c. 457, he must have from 1383 to 1387". Or, conversely, the indicator of operating profit (f. 37 s. 1 d. 2) from the account for operating result calculation (Prato, AS, D, № 377, c. 6r) was transferred to the seventh (last) position on the credit side of the same account: "And they must have for profit made on much wax of Romania in this book at c. 6, wax must give".

The "Profits and Losses" account in the examined accounting system consists of tree centre spreads – Prato, AS, D, № 377: c. 73v–74r; c. 74v–75r; 75v–76r.

The authors have constructed the dynamic lines for the formation of the totals for the complex positions of profits and losses and described the individual items for each of the six pages. The schemes reflecting the interrelation of the indicators have been reproduced for each dynamic line of accounts.

The specification of the profit indicators collected on the leaf Prato, AS, D, № 377, c. 74r, has allowed us to establish that four out of seven indicators were presented in a form of complex positions, which had been accumulated dynamically from several accounts. So, the result I, transferred from the book Prato, AS, D, № 377, c. 47v (the sum of f.1.1175, s.15d.4), was being accumulated in two books on 10 accounts and included 386 profit indicators, revealed in various sales of goods accounts. For instance, the dynamic line of the determination of profit total IV, which had been transferred from the book Prato, AS, D, № 377, c. 71r (the sum of f.8408 s.12 d.10), was being collected in two books on 27 accounts and included 1114 of the operating result indicators. Three totals on account Prato, AS, D, № 377, c. 74r are of an individual character, i.e. they contain an operational result from one certain sales account.

In the course of the study on the batches of goods sold at profit it was established:

- the sales of goods account were kept in two books (Prato, AS, D, № 357 и Prato, AS, D, № 377);
- the overall profit was accumulated in 68 accounts, 49 out of them were complex accounts;
- there were revealed the facts of profits and other incomes calculations 1868 cases;
- the total of the profit made f. 13214 s. 2 d. 6 according to the authors' calculations, but the medieval accountant had reflected the sum amounting only f. 13210 s. 12 d. 10. The discrepancy equalled f. 3 s. 9 d. 8.

Considering the batches of goods sold at loss there have been made the following conclusions:

- the sales of goods account were kept in two books: Prato, AS, D, № 357 и Prato, AS, D, № 377;
- the overall profit was reflected in 102 accounts, 30 out of which were complex accounts;
- the loss was revealed in 937 cases, i.e. half as likely as the profit;

– the sum of the calculated loss amounted f. 11492 s. 9 d. 5 according to authors, although the medieval accountant showed the loss of f. 11488 s. 3 d 10. The detected discrepancy equalled f. 3 s. 19 d. 9.

The net operating surplus totalled to f. 1 721 s. 12 d. 6 (according to the authors' calculations), when the same indicator of the medieval accountant was f. 1722 s. 7 d. 7, which is overestimated by s. 9 d. 6, than the result of the authors.

In its turn, let us pay attention to the accounts of the financial result. As it was shown above, the "Profits and Losses" account was placed on three folios. The authors did not have a chance to come across any other earlier preserved "Profits and Losses" account like that one in the archival materials as well as their descriptions in the literature. The peculiar feature of creating such accounts at least in Datini companies was that each folio was perceived as an autonomous account. The totals, calculated on the debit and the credit sides of the account, were not transferred to the next folio, but were used to balance the account. On the new folio (the continuation of an account), the balance of the closed account was transferred using the double entry method. That was the difference between the debit and credit totals.

In this case, when closing the first account on the folio Prato, AS, D, № 377, c. 73v-74r (when the debit side of the account was filled with the recordings) the debit total equalled f. 2493 s. 17 d. 6, the credit total – f. 13129 s. 18 d. 3. So, on the first folio the amount of profits calculated on the credit side was much higher than the amount of losses on the debit side. For the purpose of balancing the account, the medieval accountant wrote down the balance in the amount of f. 10636 d. 9 on the debit side of it, which he later transferred to the credit of the continuation of the account Prato, AS, D, № 377 on c. 74v-75r.

If we consider the monetary evaluation of each record on the first folio, it turns out that the loss indicators are insignificant in comparison with the profit entries. For this reason, there are 36 entries on the debit side, and only 7 on the credit side. Earlier the filling of the debit side (26 entries against 5) also caused the closure of the second account on the folio Prato, AS, D, №377, c. 74v-75r. The closing procedure was similar to the previous one: the total on the debit side was only f. 460 s. 19 d. 8, the total on the credit side – f. 2517 s. 1 d. 8. The balance of the closing account was transferred to the credit of the final account. It amounted f. 2056 s. 2.

In the final account (folio Prato, AS, D, No. 377, p. 75v-76r) the total on the debit side was f. 391 s. 5 d. 6, the total on the credit side – f. 2113 s. 12 d. 1. The accountant did not balance it by calculating for the debit side that did not add up to the credit total (f. 2113 s .12 d. 1). He kept their real values both in the debit and the credit of the account. So, only below the debit side of the account (where there was some empty space) he recorded the balance – the profit in the sum of f. 1722 s. 6 d. 7.

Francesco Datini was the sole owner of the enterprise. He did not need to distribute profits among any partners; he did not create the reserves out of profit.

The methodology for the determination of profit, despite the absence of summary indicators for profits and losses on the "Profits and Losses" account, a well-established system of the account organisation and the use of relevant accounts, allows one to gauge the high level of the accountant's professionalism. It is especially important that thanks to the efforts of Francesco Datini, the subsequent generations of the bookkeepers, of the scientists of the beginning of the 20th century (Benelli, Guasti, Livi, Lungo, Nicastro, and Bensa (Kuter et al. 2017) and, of course, Federigo Melis) managed to preserve all the account books that allow studying such an important early fragment in the history of double-entry bookkeeping.

There is only one circumstance that has disappointed the researchers, to which, in all probability, it will never be possible to get an answer. The fact is that all accounts, at the closing of which their balances were transferred either to the accounts for accumulation of the operating results or directly to the "Profits and Losses" account, are dated from 1383 to 1386. All of them have cross-references to cartas 73v–76r. The "Profits and Losses" account contains references back to the corresponding accounts. It is still not quite clear why on all the six pages of the three folios, on which the "Profits and Losses" account is placed, there is written the year 1406, which is the period of liquidation of the Datini companies in Pisa.

This is the only reason that does no let the authors state that the "Profits and Losses" account of the Datini enterprise in Pisa (1383–1386) was one of the earliest preserved examples.

4   Conclusion

The paper considers one of the earliest practical examples of a well-organized system of the determination of profit in the accounts by means of double entry method. The absence of a Trial Balance in the accounting system of the Datini company (1383–1386) does not allow it to be attributed to double entry bookkeeping. However, an equally important assumption relating to the accounting history could be put forward: the "Profits and Losses" account could have appeared earlier than the Trial Balance, but not simultaneously with it. The paper establishes a clear hierarchy of the accounts for calculating the financial result: the accounts for operating result calculation (sales accounts), operationally accumulating accounts according to the types of activity (separately for profits and losses) and the financial result accounts ("Profits and Losses" account).

The issue of the study, which is left open, is the fact that the date of the "Profits and Losses" account creation (1406) does not coincide with the dates on the accounts, the balances of which determine profit or loss (1383–1386).

The authors hope that the results of the long-term research outlined in this paper will serve as a scientific contribution to the accounting history and will prove the feasibility of the research based on real archival materials.

References

Bensa, E.: Francesco di Marco Datini: discorso detto nell'Aula maggiore del Comune di Prato il dì 21 agosto 1910. ARTI GRAFICHE CAIMO & C, Genova (1923)
Bensa, E.: Le forme primitive della polizza di carico: ricerche storiche con documenti inediti. Caimo, Genova (1925)

Bensa, E.: Francesco di Marco da Prato. Notizie, e documenti sulla mercatura italiana del secolo XIV, Treves, Milano (1928)

Besta, F.: La Ragioneria, vol. 3, 2nd edn. Rirea, Rome (1909). Facsimile Reprint, 2007

Carradori, A.: Francesco di Marco Datini, mercante pratese del sec, vol. XIV, Prato (1896)

Ceccherelli, A.: I libri di mercatura della Banca Medici e l'applicazione della partita doppia a Firenze nel secolo decimoquarto. Bemporad, Firenze (1913)

Ceccherelli, A.: Le funzioni contabili e giuridiche del bilancio delle società medievali. Riv. Ital. Ragioneria **14**(8), 371–378 (1914a)

Ceccherelli, A.: Le funzioni contabili e giuridiche del bilancio delle società medievali (Continuazione e fine). Riv. Ital. Ragioneria **14**(10), 436–444 (1914b)

Ceccherelli, A.: l linguaggio dei bilanci: formazione e interpretazione dei bilanci commerciali. Le Monnier, Firenze (1939)

Corsani, G.: I fondaci e i banchi di un mercante pratese del Trecento: Contributo alla storia della ragioneria e del commercio: da lettere e documenti inediti. La Tipografica, Prato (1922)

de Roover, R.: The development of accounting prior to Luca Pacioli according to the account-books of Medieval merchants. In: Littleton, A.C., Yamey, B.S. (eds.) Studies in the History of Accounting. Sweet & Maxwell, London (1956)

Del Lungo, I.: Francesco di Marco Datini mercante e benefattore. Giaclietti, Prato (1897)

Guasti, G.: Ser Lapo Mezzei, lettere di un notario a un mercanto del secolo XIV, Florence (1880)

Kuter, M., Gurskaya, M., Andreenkova, A., Bagdasaryan, R.: The early practices of financial statements formation in Medieval Italy. Account. Hist. J. **44**(2), 17–25 (2017)

Kuter, M., Gurskaya, M., Bagdasarian, R., Andreenkova, A.: Depreciation accounting in Francesco Datini's companies. In: 5th International Conference on Accounting, Auditing, and Taxation (ICAAT 2016) (2016). https://doi.org/10.2991/icaat-16.2016.26

Livi, G.: Dall'Archivio di Francesco Datini mercante pratese. F. Lumachi, Firenze (1910)

Melis, F.: L'archivio di un mercante e banchiere trecentesco: Francesco di Marco Datini da Prato. Moneta e Credito, VII (1954)

Martinelli, A.: The Origination and evolution of double entry bookkeeping to 1440. ProQuest Dissertations & Theses Global pg. n/a (1974)

Nicastro, S.: L'Archivio di Francesco di Francesco di Marco Datini in Prato (1914). Rocca S. Casciano, L. Cappelli

Nigro, G.: Francesco di Marco Datini: The man the merchant. Firenze University Press, Firenze (2010)

Origo, I.: The Merchant of Prato, Francesco di Marco Datini. London, J. Cape (1957)

Zerbi, T.: Le Origini della partita dopia: Gestioni aziendali e situazioni di mercato nei secoli XIV e XV. Marzorati, Milan (1952)

Information Technology in Communication

Generation Y: The Competitiveness of the Tourism Sector Based on Digital Technology

Pedro Liberato[1]([✉]), Dália Liberato[1], António Abreu[2],
Elisa Alén-González[3], and Álvaro Rocha[4]

[1] School of Hospitality and Tourism, Polytechnic Institute of Porto,
Vila do Conde, Portugal
pedrolib@esht.ipp.pt
[2] Porto Accounting and Business School, Polytechnic Institute of Porto,
Porto, Portugal
[3] Faculty of Business Sciences and Tourism, University of Vigo, Ourense, Spain
[4] University of Coimbra, Coimbra, Portugal

Abstract. In this article we present the literature review on the influence of Generation Y on tourism and its relationship with the supply of hyper personalized products. Also seeks to reflect on the changes in tourism in the digitalization era. Also known as Generation Me, this generation are digital natives, multicultural, able to change, descendants of globalization, protagonists of selfies, preferring personalized products and tailored services, which is one of the market greatest challenges. They have more rational characteristics to consume. Regarding the tourism industry, the development of technology and the internet have been an asset. The adoption of information and communication technologies has led to changes in the way of communication with the individual or institutional clients and enables the adoption of innovative business models and electronic sales channels of tourism products. It is crucial the information about demand/tourists, tourist destinations, amenities, availability, pricing, geographic information and weather, supply and transport, information about companies, intermediaries and competitors; trends in the tourism market, prices, products and tourist packages, Marketing Organization of tourist destinations, trends in the industry, size and nature of tourism, as well as policies and development plans. The role of the ICT in tourism has become an essential tool in today's world of quick information, especially on Generation's Y options.

Keywords: Generation Y · Information and communication technologies
Internet · Tourism · E-tourism

1 Introduction

To many countries, tourism plays an important role in generating revenue for the nation. Tourism is a major export industry for many countries and cities [48] and as an important and necessary industry, it is considered a sector of the economy that can benefit from the various technological resources available [26]. Despite all the political,

© Springer International Publishing AG, part of Springer Nature 2018
T. Antipova and Á. Rocha (Eds.): MosITS 2017, AISC 724, pp. 227–240, 2018.
https://doi.org/10.1007/978-3-319-74980-8_21

economic and social instability, people continue to feel like traveling. Tourism is an industry with its own characteristics, with great importance in the economy of many countries, including Portugal [33, 37], based on its high potential for generating income and employment, contributing to the increase in the gross domestic product (GDP) of each country [33, 50]. The use of Information and Communication Technologies can provide competitive advantages in promotion, as well as strengthen the strategies and operations of the tourism industry [12, 33].

[46] evidences that tourism adds a variety of services that can benefit from Information and Communication Technologies such as: accommodation services that include hotels, apartments, residences and villages; catering services, including restaurants and cafés; transport services including trains, buses, taxis and aviation companies; tourist support services such as insurance companies and exchange offices; recreational services such as sports centers, swimming pools, marinas and golf courses; attractions services such as museums, parks, gardens, monuments and conference centers; entertainment services including casinos, cinemas, clubs and theaters. Also [13, 33, 47] consider crucial information about demand/tourists, tourist destinations, amenities, availability, pricing, geographic information and weather, supply and transport, information about companies, intermediaries and competitors; trends in the tourism market, amenities, availability, prices, package tours, Marketing Organization of tourist destinations, trends in the industry, size and nature of tourism, as well as policies and development plans.

On the other hand, [40] highlights the link between information and communication technologies and the tourist activity, saying that "(…) contrary to what may seem at first sight, tourism and ICT can be regarded as two sides of the same coin". [33] point out that a diversity of suppliers throughout the supply chain are now able to form a direct connection with customers through digital platforms, providing visitors more familiar with them, the ability to create personalized trips. If we make an historical analysis about tourism development, it is closely linked to technological evolution. Thus, from [40] perspective, communication and information tools are essential to tourism activity, and it is fundamental that tourism service providers consider the contemporary tourism "ecosystem", formulating strategies for the sector and for tourism destinations [33], because, while some tourists prefer the convenience of group trips organized in packagings, others hope to get more interactive and personalized services [24].

Tourism, nowadays, has become a very frequent leisure activity especially for the so-called Generation Y, so this market segment has been emphasizing itself as a very important niche and of great interest for the tourism sector [4]. Have consumer features little decipherable, because this generation has grown up in an increasingly globalized world, in the digital age and where knowledge has undergone radical changes and abrupt, especially with the use of technology that allows these young people ease of contact and relationships with people from different countries [4]. The tourist from generation Y is considered as a very flexible tourist at the time of travel, for dates and destinations. Has not set a standard, or have specific preferences. Are influenced by behaviors in group and reserve time for fun. Thus, some tourist destinations offer alternative activities according to the consumption habits, as the practice of sports, enjoy parties/events, conduct meetings with other young people (e.g. pub crawl), try new cuisines, discover unknown cultures, exchange of experience in large scale events, among others. We can thus say that there is no specific activity that defines what actually this generation seeks, or do while they travel. However, according to The World Youth Student and Educational Travel Confederation (2014), this generation, while traveling, want to relate to the maximum with the residents, want to live an experience of everyday life in the destination and increase their knowledge [4]. The most popular applications include: TripAdvisor, Airbnb, Skyscanner, Booking, among others, that allows to view comments from other tourists about different places, what is crucial for this generation, that is, the sharing of feedback for a reliable effect. The tourism industry shows a special interest in the change of behaviors, between generations, with the objective of adapting its offer, for each segment of demand, through studies that reveal behavioral characteristics of the tourists. Nowadays, they travel more and not only according to seasonality; they are exploring more and more different destinations (not returning to a destination already visited); have higher purchasing power; they do most of the online booking; require more information available and look for new experiences. Considered a new way of life, tourism destinations and Destination Management Organizations (DMO's) and all its components need to perceive these characteristics, being able to adapt their methods of marketing products and services [44].

This analysis seeks to understand the role of Information and Communication Technologies and the Internet in most of the activities associated with the tourism sector.

2   Technological Development: The Internet

As [1, 33, 34] points out, Information and Communication Technologies have a strong influence on people's daily lives. The current society is somehow linked to a concept of knowledge-based economy, a "(…) economy that is focused on knowledge and information as bases of production, productivity and competitiveness" [17]. Essential, today, is to carry out a correct use, research, storage and processing of information, making it essential to learn how to deal with technology and with all the information that is available. People have to follow properly the innovations towards effective integration into the labor market and services. On the other hand, and as [50] concern, wealth creation must be the relationship between individuals and institutions, as well as the ability to manage the existing means and resources in the territory. Technological development has contributed to several changes that occur in society and in people's behaviors, particularly in how we communicate, how people behave and interact in society [18], as well as in the way we seek products, services and information [1, 14–16, 24, 33].

[36] refer that information and communication technologies have not only changed the way people conduct their activities, but also the mobilization of material and immaterial resources, the way in which generates wealth and how negotiated opportunities are created and expanded. [47, 49] emphasize that "the internet, today, represents one of the main information and communication technologies. This new tool works, through thousands of interconnected computers, around the world, enabling data exchange and information provided in a large network". To [8] the internet, with

regard, specifically, to tourism, has changed profoundly the way tourists access to information, plan their trips, make reservations, and share their travel experiences. The internet and other interactive technologies, in addition to causing changes in the behavior of people drive changes in the market, in particular the tourism market, enabling the global distribution of tourist services [45, 55]. Technologies use makes the market more competitive and more accessible to the user [56].

The World Tourism Organisation [58], assumes that the internet and tourism are ideal partners. The internet meets the needs of the tourists of the 21st century, increasingly demanding, informed and sophisticated. The internet reduces geographical distances between tourism companies and customers, which may contribute to greater flexibility, mobility, elasticity and efficiency in business.

As [3] says, the Internet enables a wide social network (virtual), linking the various subjects by the most diverse forms, astonishing speed and in most cases, a synchronous interface, giving a new concept of social interaction. By reducing geographical distances between tourism companies and customers, creates greater flexibility, mobility, elasticity and efficiency in business. The Internet is a strategic resource and [53] "(…) it will be enough for the agents and promoters of the hotel sector to become aware that the Internet, being a useful window for promotion and dissemination, is an ideal space for conducting business. In order to take full advantage of this communication tool, agents and promoters in the hotel sector must centralize their core business by making proposals, on the internet, based on flexibility models, speed, utility and imagination. [19] point out that companies that provide tourism services, when setting up their own websites have the possibility to establish direct contacts with consumers, which will help to increase sales. On the other hand, [38] emphasize that the Internet allows tourism providers to be in the same place as their clients or potential clients, and to understand their attitudes, needs, interests, choices and requirements. Technologies promote the exchange of information which is essential for tourism [3, 33, 46, 47]. [9] refers that tourism activity generates a significant amount of information that has a strategic importance and value to the business. This means that the information must be treated as a strategic element of organizational/institutional planning. Using the internet, tourists have immediate access to useful information, varied destinations, and the possibility of making reservations in an easier and faster way. Particular attention should be given to changes in market needs, triggered by technological innovations [14] and, especially to a new market resource, the mobility and ubiquity allowed by the spread of smartphones and the emergence of QR codes that contextualize the mobile applications and services and renew the discussion about the importance of the destination strategy [1, 10, 25, 29, 33, 37]

2.1 The Emergence of e-tourism

Tourism companies will gain competitive advantage if they are able to maximize their profits by improving their services in order to achieve a greater degree of consumer satisfaction [11, 12, 33]. In this technological context, the destination is understood to be a variety of individual products and opportunities for experiences. Regarding to the tourism industry, the development of technology and the internet have been an asset, since according to [15] contributed to several changes that seek to improve the offer in terms of services. The Internet and other interactive multimedia platforms contribute to the promotion of tourism by enhancing transformations in the tourism industry structure [12, 13]. Experts at the 19th World Travel Monitor Forum (ITB Berlin, 2011) point out that tourism needs to be online, beyond traditional marketing strategies. They also add that people worldwide prefer to use the internet to book their trips. In the 20th Monitor Forum WorldTravel (ITB Berlin, 2012) concluded that people seek to take advantage of the latest technologies in terms of information and to purchase products and services. It is also noted that the purchase of travel and other tourism products is extremely popular online. The report *Future Traveller Tribes 2030, Understanding Tomorrow Traveller*, published by Amadeus Traveller Trend Observatory in April 2015, refers that travel trends, in the next few years will be determined primarily by the intensive use of technologies, social and cultural criteria, the convenience in travel management and trip, the lack of time, and luxury.

For many tourists, technology represents an opportunity to actively participate in the destination activities and to take part personally in the construction of their own experience [45]. Likewise, they place special emphasis on sharing their experience with other tourists and residents, and are therefore willing to activate conversation processes through social media with the destination using electronic devices [16], with their family, friends or anonymous users [10, 42]. In this sense, it has been shown that the most valued experiences are those co-created with tourists and supported by high levels of technology [33, 52]. As argued by [52], ICTs are extremely useful because they facilitate encounters between tourists and the destination, and improve the experiential process in time and space.

Tourism companies, feeling the need to adapt to technological development, included the technologies in their business processes, and thus emerged the e-Tourism. According to [11] e-Tourism represents all aspects of tourism that involve and promote the integration of information and communication technologies, revolutionizing the strategic relations of tourism organizations and all of its stakeholders [15]. The concept of e-Tourism encloses all functions of business such as e-commerce, email marketing, electronic production, as well as the electronic strategy, electronic planning and management for all sectors in the tourism industry [11], grouping, also according to the same authors, three main areas: business management, information systems and tourism management. [13] highlight that e-tourism is a result of the scanning of all processes and the value chain of the tourism sector, in particular travel, hotel and restaurant management. E-tourism, is also the result of the fact that we live in the age of wireless communication and that tourists use their mobile devices with internet access before, during and after their trips [32].

[20] refer to the electronic or virtual agency that allows users to [20] "search, plan and make all the reservations they wish from a computer terminal, allowing the issuing airline tickets, reservations in hotels, rent a car, choice of cruises, among others, with all partners associated with such virtual agencies", whereas [27] consider very important the existence of applications for smartphones that offer tourists a wide variety of services such as audio guides, road maps, interactive info on transport, tariffs, cultural agendas, among others, as a strategy to support the decision on the choice of destination.

2.2   Big Data

Companies face today, many challenges, such as the storage and treatment of numerous and varied data generated in the course of its activities. According to [29], innovation, business model transformation, globalization and personalization services, observed nowadays, has contributed to the increase of data generated. Globalization has not only trade and even the way of working, but also the variety of data format. Organizations need to be effective, provide to their client a quickly and efficiently service. To do this they need to understand and meet the needs of its customers. In addition, companies must take into account new data sources generated by social networks, mobile devices and sensors [29]. However, they need and access a lot of information, but they do not always get value from the same face of their large quantity and lack of structuring [59].

It is in this context that Big Data allows the access and analysis of large amounts of data. Thus, analytical intelligence makes it possible for companies to become more competitive [21]. Big Data is not a "thing", but a dynamic activity that crosses many IT frontiers" [25], constituting itself as a large data set, whose challenge is storage, research, sharing the visualization and analysis of the information that is generated, understanding [57] Big Data as a collection of complex and voluminous databases that do not allow simple operations effectively with management systems of traditional database. For [57], Big Data Analytics is a technologic strategy that allows companies to more intensely and accurately perceive customers by analyzing patterns and correlations, enabling tourism companies to gain more advantages and become more competitive. Cloud computing enables small and medium businesses to implement Big Data technology.

By the end of the 20th century and the beginning of the 21st century, and as a result of the influence of the Internet and e-commerce, a great quantity and variety of data was instantly produced as users transmitted it through the World Wide Web. [22], point out that the amount of data that may be of interest and can be used by both business and people increases every day.

Faced with such a large amount of data, the most complex becomes its analysis and understanding. [36] or [39] report that data analysis will allow a better understanding of customers, markets, competitors, products, the market environment, the impact of technologies or suppliers. We now benefit from other transformations resulting from the ubiquity of mobile devices, the use of cloud computing and the connection of various everyday devices, dubbed the "internet of things" [36], which centralize data production for a continuous flow.

2.3   Internet of Things (IoT)

With the increasing use of smartphones and tablets, the number of devices connected to the Internet also enlarged. The type of communication known as the Internet of things appears linked to a worldwide network of interconnected objects [28]. The Internet of Things (IoT) covers the use of networks, sensors and cloud computing [7, 30, 31, 41] allowing the link between physical objects and computers with the internet. According to [6] the Internet of things allows the development of a large number of applications in various areas and environments, integrating objects such as mobile phones, sensors, and other devices that interact with each other. In the Internet of things the objects work in smart spaces, using intelligent interfaces to relate and communicate with various environments [51]. These concepts are formed by ubiquitous computing, pervasive computing and intelligent environments [23]. [56, 57] also agree that the ubiquitous computing has become invisible to the end user computing, and pervasive computing suggests that user's access to information and computer resources regardless of the location or device used. [23] consider that in smart environments, devices can interact with the processes. In turn, [51] report that in the environment of the Internet of things, objects acquire naturally three features: intelligence, connectivity and interaction.

The development of new technologies will enable the use of smart devices in everyday life [31, 51]. With the Internet of Things, numerous objects will be linked together, with increased traffic volume, and data storage capacity [51].

2.4 Virtual Reality and Augmented Reality

Virtual reality consists of a three-dimensional, computer generated medium that allows the user to see, interact and manifest in an environment outside reality. Depending on the interactivity provided may be immersive (based on use, for example helmets) or non-immersive (based on the use of monitors). Got a lot of notoriety with Second Life.

Augmented reality is a technology derived from virtual reality that consists of superimposing digital information to real-world images, that is, it does not completely emerge in a virtual world where it cannot see what is around it: it is a supplement to the reality and not a substitution of it. Can also be seen as a middle ground between Virtual reality (fully synthetic) and Telepresence (quite real), according to [41]. [5] states that the main difference of virtual reality with regard to augmented reality lies in the fact that in augmented reality we have the possibility to visualize objects and graphics in a real environment while also allowing virtual reality although in an environment separated from reality.

Given the great potential of this technology in several areas and tourism in particular, there is now a large proliferation of applications for mobile devices (Apps), with augmented reality, applied in museums, monuments, galleries, open spaces and tourism attractions in general, where objects can be "augmented" and complemented in real time with diverse information (text, images, three-dimensional animations, audio or video).

2.5 Location-Based Services (LBS)

The user's location-based services (Location Based Services-LBS) use GPS technology and enable the development of applications that allow to implement new models of mobility. Are example, traffic management systems, navigation systems and information to the user on the move. The same concept applied with a proximity criterion refers to the beacon. The Beacon is a small device that connects with other electronic devices and POS (point-of-sale system). In the same physical space, for example a shop, such may be connected with a payment system (e.g. Paypal). Thus, an individual with a mobile device, for example a smartphone with Bluetooth, will receive notifications as long as it is in the range of the beacon.

2.6   Generation Y in Tourism

This generation aims at their professional and personal fulfillment, thus enjoying the pleasures of life and especially like to make tourism as a recreational activity, the most frequent type of tourism for this segment is the adventure, the discovery of the new, interconnected with shopping tourism; they also try to make many of their trips outside of their country of residence, and live the philosophy You Only Live Once (yolo) in your life and in your travels. This philosophy consists in living each moment as if it were the last [4]. Given this marketing environment, researchers have a great challenge, that is not only to know their consumer behavior, but also to focus on these the best strategies of consumption, since according to the same authors [4], the impact of this market segment is very significant for the tourism industry as it represents approximately 190 million international trips per year and approximately 165 billion dollars according to the World Tourism Organization (UNWTO) in 2014.

According to the World Stats, 42% of the population uses the Internet, while in 1995 it was less than 1%. Based on this growth, companies have begun to view the Internet as a major and central element in their marketing and communication campaigns [34].

The emergence of Web 2.0 was a turning point in the Internet world, forcing behavioral market changes to stay in the competition. With the free and public exchange of opinions and experiences acquired in the market, consumers have taken a leading position in the supply and demand market.

In a short-term, the supply side must be aware of the importance of the Internet and how it has affected consumer behavior. In 30 years the radio reached 50 million listeners; the television took 13 years to have 50 million spectators; the Internet took only 4 years to reach 50 million users; Facebook reached this number in 18 months [34]. Keeping in mind the importance of this new type of communication, companies are converting their marketing methods, adapting them to this new reality, in order to more effectively and quickly reach the market demand.

The main business organizations today have official pages on the Internet or in social networks, causing a greater impact on their customers, while passing a competitive and innovative image. Business marketing has gone from a transactional aspect to a relational aspect. Relational Marketing provides companies with the means to establish relationships, networks and interactions with different stakeholders [54]. This relational approach puts customers at the center of the whole process; it is necessary to identify what has value for them, capture them and have the ability to continue creating new values, to maintain and retain the customer. Here, the supplier/customer interaction becomes the crux, rather than the product [34].

In this way, the Internet is used as a digital marketing tool in tourism, presenting advantages such as cost reduction, speed, accessibility, fast and effective communication of a large volume of information, possibility to customize information to each user [36]. Generation Y uses the Internet in several areas, with 37% of purchases made in connection with travel and accommodation, a very significant value for the tourism sector, which is second only to total online purchases [18, 20, 33].

According to [44] there is greater contact between tourist destinations and visitors through social networks, based on low cost and efficiency. Currently social networks are considered a competitive element for the entrance of the global tourist market, because if managed well it can be distinguished by the difference. However, there is care to be taken in the organization of the information and the way it is published. This has to be appealing and creative, as more and more these networks are loaded with information making it difficult to originality and attraction.

According to [12, 16, 19], social networks can be divided into four types:

- Relational Networks - creation of new virtual groups and communities; easy connection between people (ex: Facebook);
- Sharing Networks - dissemination of photos, videos and varied information (e.g. Youtube);
- Publication networks - through articles and posts provide a quick update and provision of information (e.g. blogs);
- Microblogging Networks - Short information, which can be accessed by all or only by a restricted group (defined by the user), through photos or texts (e.g. Twitter).

As we can see, the Internet is always present, and is decisive in the purchase decision, in the dissemination of the experience and in the "word-of-mouth" about the destination visited. This last phase is crucial for the tourism destinations websites, and online agencies. The Internet has a pivotal role in the perception, construction of the image and choice of a tourist destination.

Regarding tourism destination promotion, today, the image, allied to the new forms of communication of the Generation Y becomes very important. When people travel, they need to share what they see, what they eat, where they are, in real time, sharing images in social networks. Virtual tourism communities emerged in the late 1990s. Today, we have web forums, videos, blogs, social networks (some of them specifically geared towards tourism) and virtual agencies, which provide data about people in a fast, efficient and persuasive way, more diverse tourism destinations, even being an excellent diffusion of small businesses, that otherwise would not be able to reach such a wide public [56, 57].

The social networks in the tourism and the behavior of the consumer "online", are currently obligatory items in the management of the product/tourist service. These images, shared by people the potential tourist knows (which gives credibility to the information transmitted), can advertise destinations that are not even present in traditional media, brochures or tourist magazines. In addition, this information is always available on the Internet, and can be consulted at any time, and shared by any user [44].

Speed is crucial; the Generation Y wants information now, and being difficult to find online, simply look for another source of information more accessible, in the well-known websites like Google, Tripadvisor, etc. For the operator this is ungrateful, as it has little control over the information there. Given this evidence, they must provide up-to-date, organized, and easy-to-find information. Generation Y often uses the expression "there is an app for everything". And tour operators know that the mobile version of the information is very important. Many times the trip is planned in a meeting between friends, in a bar, on the beach. According to an Expedia survey, 49% of Generation Y uses the smartphone to plan a trip, 40% to share, during the experience and 35% to make reservations [13, 32, 33].

The personalized relationship with the client, digital technology and the internet world allow the collection of information about each user, creating individualized "packages" of information. By receiving "feedback" from tourists from around the world and at the same time having the possibility to analyze the type of search they make as well as the reservations that prevail, tourism companies build giant databases through racking of mobile data, fundamental for the construction of a "tailor made" offer of the customer [44]. This tool opens new opportunities to adapt this concept to the concept of hyperpersonalized products/services, being an offer that the generation Y, more and more, seeks. As mentioned previously, this generation of tourists when traveling is differentiated by their preferences, namely, meeting new people and enjoying experiences, rather than material purchases. Sites such as Airbnb, Home-Away, EatWith and Dopios have met the needs of Generation Y, making it innovative and enterprising at the same time. There was a specific set of needs of a new generation of consumers, which brought radical changes to the traditional market [35]. Tourism distribution began to be made by digital means allowing customers to control the search for information, obtaining discounts and usufruct of the absence of the rate that tour operators and travel agencies charge. However, this new distribution of the sector also has its less positive points, such as: the loss of time needed for research, lack of specialized advice and too much information available. With this, operators and agencies also lose their relationship with suppliers and customers, and consequently lose profits. For suppliers, the electronic distribution is an advantage, since they no longer pay commissions to intermediaries [46].

The volume of online purchases of tourist services has grown significantly in recent years, especially the airlines, which is causing a change in marketing strategy. An hyperpersonalized product presents advantages in the perspective of the consumer, financial benefits, saving time in the reservation and purchase of products, whose correspond to their desires and needs [46]. With the use of new technologies, it is possible for tourists to choose personalized products at a competitive price, allowing the creation of individual packages that meet the tastes of each one. It is a great advantage for tourist demand as it satisfies the needs of the consumer and has added value, eventually reducing costs, allowing purchases of other complementary products during the trip [46].

3   Final Considerations

Agents that use the internet become more competitive because they are chosen by users who prefer to opt for destinations, make hotel reservations more independently, faster and cheaper. The pertinence of the subject addressed results from its recent expansion in the international market. This analysis allows a better understanding of the promotion in the tourism sector of the future, adapting to potential clients in transformation. The way this generation and the generation that follows act, causes an obligatory change in the search for the satisfaction of their needs and expectations.

Given the overall context of promoting use of technologies in search of information about destinations and tourism enterprises, it will be decisive for the tourism sector management, investments in the professionalization of human resources with special emphasis on the enhancement of technological skills; the continuous appreciation of the mark – official product, official portal, for the submission of quality companies in the region, with a permanent concern for transparency regarding information provided to the tourist/visitor.

A shared vision should be developed between the citizen who lives in the city, the tourist that visits it, destination management, and the different stakeholders, in order to add value to the citizen and the visitor. Destination Management Organizations (DMO's) must provide citizens and tourists with a collaborative platform that allows bi-directional communication between the public administration and citizens or tourists/visitors [33]. In order to value the tourist experience at the destination, free internet should be promoted in public and private places, associated with products that incorporate the use of technology and digital media in a context of diversified offer [33].

In the tourism industry, this change of meanings makes it essential to adapt its media and marketing, to obtain more positive results and to be able to compete in this new market trend [11–13, 33]. Tourism marketing will have to establish a partnership between digital marketing and relational marketing, being more innovative and inter-active, in order to survive in an increasingly competitive market, characterized by changes in the expectations of tourist's vis-a-vis tourism destinations.

Generation Y has power over the tourist market, power of digital knowledge, having created a new way of traveling, by its own co-creation. They seek to be innovative and enterprising, even on vacation, that is, they prefer to create their own package instead of choosing one already set by a professional in the area, which in the future may cause problems for the maintenance of some enterprises if no forms of innovation and adaptation to this new trend are found. The temporal validity is in decline, given that this new digital age is in constant movement, that is, what today is new and interesting, tomorrow may not be, and is still one of the main challenges for marketing strategies.

In this sense, customized products and services arise, since it is not enough to have control over part of what they consume/acquire, but rather the need for total power in the decision and selection of any item related to their trip, in the case of tourism. In a way, the new technologies and the new media have made this change possible by allowing them to be able to follow the trends through the tracking of information of the mobile devices of each user, constituting a tool for monitored control.

References

1. Abreu, A., Rocha, Á., Carvalho, J.V., Cota, M.: The electronic booklet on teaching-learning process: teacher vision and parents of students in primary and secondary education. Telematics Inform. **34**(6), 861–867 (2016)
2. Amadeus Traveller Trend Observatory: Future Traveller Tribes 2030: Understanding Tomorrow's Travel. (2015). http://www.amadeus.com/documents/future-traveller-tribes-2030/travel-report-future-traveller-tribes-2030.pdf
3. Anjos, E., Souza, F., Ramos, K.: Novas Tecnologias e Turismo: um estudo do site Vai Brasil. Caderno Virtual de Turismo. vol. 6(4), (2006)
4. Astudillo, P., Mendoza, C.: Tendencias de consumo turístico de los Millennials en la ciudad de Ibarra. Ecos de la Academia 4(14), (2016). E-ISSN 2550-6889

5. Atzori, L., Iera, A., Morabito, G.: The internet of things: a survey. Comput. Netw. **54**(15), 2787–2805 (2010)
6. Avelar, E., et al.: Arquitetura de Comunicação para Cidades Inteligentes: Uma proposta heterogénea, extensível e de baixo custo. Universidade Federal de Pernambuco (UFPE), Recife (2010)
7. Azuma, R.: A survey of augmented reality. Teleop. Virtual Env. **6**(4), 355–385 (1997)
8. Bilgihan, A., Barreda, A., Okumus, F., Nusair, K.: Consumer perception of knowledge-sharing in travel-related online social networks. Tour. Manag. **52**(2), 287–296 (2016)
9. Bissoli, M.: Planejamento Turístico Municipal com Suporte em Sistemas de Informação. Futura, São Paulo (1999)
10. Brejla, P., Gilbert, D.: An exploratory use of web content analysis to understand cruise tourism services. Int. J. Tourism Res. **16**(2), 157–168 (2014)
11. Buhalis, D.: eTourism: Information Technology for Strategic Tourism Management. Pearson (Financial Times/Prentice Hall), London (2003)
12. Buhalis, D.: Information technology for small and medium-sized tourism enterprises. In: Keller, P. Bieger, T., (eds): The future of small and medium sized enterprises in tourism, AIEST congress 2004, Jordan, Editions AIEST, St-Gallen, Switzerland, pp. 235–258, ISBN 3952172359 (2004a)
13. Buhalis, D., Flouri, E.: Wireless technologies for tourism destinations, In: Frew, A., (eds): Information and Communications Technologies in Tourism, ENTER 2004 Proceedings, Springer-Verlag, Wien, ISBN 3211206698, pp. 27–38, (2004b)
14. Buhalis, D., Law, R.: Progress in tourism management: Twenty years on and 10 years after the internet. The state of eTourism research. Tour. Manag. **29**(4), 609–623 (2008)
15. Buhalis, D., Jun, S.: eTourism. Good Fellow Publishers Limited, Oxford (2011)
16. Buonincontri, P., Micera, R.: The Experience Co-Creation in Smart Tourism Destinations: a Multiple Case Analysis of European Destinations. J. Inf. Technol. Tourism **16**, 285–315 (2016)
17. Castells, M.: La Ciudad de la nueva economía, La Factoría, 12 (2000)
18. Castells, M.: A sociedade em rede. Editora Paz e terra S/A, São Paulo (2002)
19. Costa, J.; Águas, P.; Rita, P.: Tendências Internacionais em Turismo, 2.ª Ed., Lidel: Lisboa (2004)
20. Coutinho, L., Sarti, F.: Nota Técnica Parcial: tecnologia da informação aplicada ao turismo. Centro de Gestão e Estudos Estratégicos. Ministério do Turismo (2007)
21. Davenport, T., Barth, P., Bean, R.: How Big Data is Different. Harvard Business Review (2012)
22. Demchenko. Y., Grosso, P., De Laat, C., Membrey, P.: Addressing big data issues in scientific data infrastructure. In: International Conference on Collaboration Technologies and Systems (CTS). IEEE (2013)
23. Dohr, A., Modre-Osprian, R., Drobics, M., Hayn., D., Schreier, G.: The Internet of Things for ambient assisted living. In: Seventh International Conference on Information Technology. IEEE (2010)
24. Dwyer, L., Forsyth, P., Rao, P.: The price competitiveness of travel and tourism: a comparison of 19 destinations. Tour. Manag. **21**(1), 9–22 (2000)
25. Gantz, W.: Reflections on communication and sport: on fanship and social relationships. Commun. Sport **1**(2), 176–187 (2012)
26. García-Crespo, A., Chamizo, J., Rivera, I., Mencke, M., Colomo-Palacios, R., Gómez-Berbís, J.M.: SPETA: social pervasive e-tourism advisor. Telematics Inform. **26**, 306–315 (2009)

27. Ji Hoon, P., Cheolhan, L., Changsok, Y., Yoonjae, N.: An analysis of the utilization of Facebook by local Korean, governments for tourism development and the network of smart tourism ecosystem. Int. J. Inf. Manag. **36**(6), 1320–1327 (2016)
28. Jiang, Y., Zhang, L., Wang, L.: Wireless sensor networks and the internet of things. International Journal of Distributed Sensor Networks. Research Center for Mobile Computing, Tsinghua University, Institute of Microelectronics, 1–7, Tsinghua University, China, (2016)
29. Krishnan, K.: Data Warehousing in the Age of Big Data. Newnes, Oxford (2013)
30. Kurose, J., Atzori, L., Lera, A., Morabito, G.: The internet of things: a survey. Comput. Netw. **54**(15), 2787–2805 (2010)
31. Kurose, J.F., Ross, K.W.: Redes de computadores e a Internet: uma abordagem top-down, 5th edn. Addison Wesley, São Paulo (2010)
32. Langelund, S.: Mobile travel. Tour. Hosp. Res. **7**, 284–286 (2007)
33. Liberato, P., Alén, E., Liberato, D.: A importância da tecnologia num destino turístico inteligente: o caso do Porto. In: Proceedings of XIX Congreso AECIT, Tenerife (2016). ISBN 978-84-617-5964-4
34. Madni, G.R.: Consumer's behavior and effectiveness of social media. Global Journal of Management and Business Research: e Marketing (2014)
35. Manolis, P., Pastras, P.: The new sharing economy: is this the end of tourism as we know it. Turismo 15, Tourism Trends Review, 77–80 (2015)
36. Marchand, D., Peppard, J.: Why IT Fumbles Analytics. Harvard Business Review (2013)
37. Maurer, C., Schaich, S.: Online Customer Reviews Used as Complaint Management Tool. Inf. Commun. Technol. Tourism **2011**, 499–511 (2011)
38. Maurer, C., Wiegmann, R.: Effectiveness of advertising on social network sites: a case study on facebook. In: Law, R., Fuchs, M., Ricci, F. (eds.) Information and Communication Technologies in Tourism 2011. Springer, Vienna (2011)
39. Mayer-Schönberger, V., Cukier, K.: Big Data: A Revolution That Will Transform How We Live, Work, and Think, (2014). ISBN-10: 0544227751
40. Mendonça, F.: A Promoção de Destinos Turísticos na Internet – O Algarve e os seus Concorrentes – Uma análise comparativa, Dissertação de Mestrado em Gestão de Sistema de Informação. Universidade de Évora, Évora (2002)
41. Milgram, P.: A taxonomy of mixed reality visual displays. In: IEICE Transactions on Information Systems, E77-D (1994)
42. Neuhofer, B., Buhalis, D., Ladkin, A.: Conceptualising technology enhanced destination experiences. J. Destin. Mark. Manag. **1**(1), 36–46 (2012)
43. OMT: E-Business for Tourism: Practical Guidelines for Tourism Destinations and business, Ed. OMT: Madrid (2001)
44. Pinto, M.: A influência das redes sociais na perceção e escolha de um destino turístico na Geração Y. Instituto Universitário de Lisboa, Dissertação de Mestrado (2016)
45. Prebensen, N., Vitterson, J., Dahl, T.: Value co-creation significance of tourist resources. Ann. Tour. Res. **42**, 240–261 (2013)
46. Ramos, C., Rodrigues, P.M., Perna, F.: Sistemas e Tecnologias de Informação no Setor do Turismo. J. Tour. Dev. **12**, 21–32 (2009)
47. Ramos, C.M.: Sistemas de informação para a gestão turística. Tour. Manag. Stud. **6**, 107–116 (2010)
48. Romão, J., Leeuwen, E., Neuts, B., Nijkamp, P.: Tourist loyalty and urban e-services: a comparison of behavioral impacts in Leipzig and Amsterdam. J. Urban Technol. **22**(2), 85–101 (2015)
49. Sant'anna, A., Jardim, G.: Turismo on-line: oportunidades e desafios em um novo cenário profissional. Observatório de Inovação do Turismo. Revista Acadêmica. 2(3), (2007)

50. Serrano, A., Gonçalves, F., Neto, P.: Cidades e Territórios do Conhecimento – Um novo referencial para a competitividade. Edições Sílabo, Lisboa (2005)
51. Tan, L., Wang, N.: Future internet: the internet of things. In: 3rd International Conference on Advanced Computer Theory and Engineering, vol. 5, pp. 376–380 (2010)
52. Tussyadiah, L., Fesenmaier, D.: Mediating tourist experiences access to places via shared videos. Ann. J. Res. **36**(1), 24–40 (2009)
53. Vector21.Com.: A Hotelaria Portuguesa na Internet – 2° Relatório Portugal Insite/Plano21. Com. http://www.vector21.com/pd/estudosmercado/[3-11-2008] (2008)
54. Vicentin, I., Hoppen, N.: Tecnologia aplicada aos negócios de Turismo no Brasil. Turismo Visão e Ação, 4, 11ª Ed, 79–105 (2002)
55. Vicentini, A., Ferreira, G., Lorenzi, F., Augustin, I.: Arquitetura de um sistema de informação pervasivo para auxílio às atividades clínicas. Revista Brasileira de Computação Aplicada. Passo Fundo, 2(2), 69–80 (2010)
56. Videira, R.A.A.: Fatores potenciadores do marketing relacional online no turismo: uma análise a agências 100% virtuais. Dissertação de mestrado em Marketing. Universidade da Beira Inteira, Ciências Sociais e Humanas. (2010)
57. Vieira, M., Figueiredo, J., Liberatti, G., Viebrantz, A.: Bancos de Dados NoSQL: Conceitos, Ferramentas, Linguagens e Estudos de Casos no Contexto de Big Data. Retrieved from http://data.ime.usp.br/sbbd2012/artigos/pdfs/sbbd_min_01.pdf (2012)
58. World tourism organisation, tourism 2020 vision. In: Global Forecast and Profiles of market Segments, vol. 7, p. 123 (2001)
59. Zikopoulos, P.C., Eaton, C., Zikopoulos, P.: Understanding Big Data: Analytics for enterprise Class Hadoop and Streaming Data. McGraw-Hill professional, New York (2011)

Information Systems Infrastructure – Importance of Robustness

Seppo Sirkemaa[✉]

University of Turku, Turku, Finland
seppo.sirkemaa@utu.fi

Abstract. Information systems have a critical role in organizations. Practically all companies and organizations depend on information systems and their operations. Therefore, robustness and reliability are key issues in information systems and information systems infrastructures. Trustworthiness of technologies and systems is important, meaning that these need to be the goals of information systems management and development. In this article, we look at development of robustness of organization's information systems infrastructure. The goal is to find the key issues that affect the infrastructure, impacting reliability and continuity of systems and technologies in the organization. Information systems need to be developed so that robustness and flexibility could be built into information systems and infrastructures. The goal is to develop systems that enable rather than restrict, so that systems act as a robust basis for activities and operations, and make it possible to reach for operative and strategic goals.

Keywords: Information systems · Development · Management
Reliability

1 Introduction

The role of information systems is critical in organizations. In all types of public and private organizations, independently of field and operations do information systems, networks and technology have a key role in organization's operations. Practically all organizations use computers in their activities in some degree, even in the smallest businesses technology has typically a role in administrative processes, marketing or in keeping in contact with the customers.

Today organizations rely on technology in high degree, and even minor technical malfunctions can impact operations and have a major impact on the whole organization. In many cases operations halt altogether when technology is not working as expected. It seems that today there are even no plan B alternatives for activities. For example, in a store with computer-based sales systems it is better to wait the system to be online again to do anything else. This example indicates the importance and critical role of information systems, they are the basis of operations and without robust, reliable and smoothly operating information systems business activities and operations become impossible.

© Springer International Publishing AG, part of Springer Nature 2018
T. Antipova and Á. Rocha (Eds.): MosITS 2017, AISC 724, pp. 241–244, 2018.
https://doi.org/10.1007/978-3-319-74980-8_22

In this article we study information systems management. The emphasis is on information systems infrastructure, which is here understood as systems and technologies, which are shared among different users and functions across the organization. Infrastructure is the basic mortar that other business systems and processes use for various purposes. Information systems infrastructure is the cornerstone of operations, and therefore it is important that it is reliable, robust, but can still be modified to support different processes and purposes. This is increasingly important as new technologies, systems and innovations are being introduced. Consequently, developers of information systems are challenged to develop their systems and solutions to keep up with the digital transformation in the economy [1].

Information systems management is about understanding the critical systems and components in the organization's information systems infrastructure. Resources are typically scarce, so there is a need to focus them on key areas in the systems and infrastructures. The question here is what are these issues?

We suggest looking at activities which increase reliability, and make it possible to continue as smoothly as possible should a disaster take place. Protecting systems and data from unauthorized access should be planned. Developing recovery procedures so that continuing operations after possible interruptions are also important. Management commitment and user awareness are critical ingredients in successful information systems development. Therefore, preventive actions which increase user awareness and skills are important.

2   Towards Robustness

Information systems management involves management of a wide range of technologies and systems. The goal is to develop systems and solutions that serve various business processes and activities. Information systems are in many cases part of manufacturing, logistics, administrative processes or customer service. Their role is often critical, as interruptions can have major impact on the organization, processes and activities in different units and departments. Clearly, there is a need to develop robust and reliable information systems and information systems infrastructures [2, 3]. This is also referred to as trustworthiness of information systems.

Information systems management is about development and maintenance activities. In many activities there are both development and maintenance-related aspects. For example, when troubleshooting an application that is frequently malfunctioning the situation may motivate to finding better solutions to this particular system. In daily life of IT department there are numerous activities, mostly about troubles and ad-hoc repairs to existing systems and technologies. It tends to be an overwhelming battle against all kinds of problems, with limited resources. The question is what should be done to have a system that is reliable and robust?

Understanding the importance of robust information systems infrastructure is important. It helps in focusing resources on those areas which strengthen the infrastructure, increase the reliability of the systems, and help users better utilize the systems. Having a better, more robust and reliable infrastructure and system is important, and so resources need to be focused on key issues in information systems infrastructure.

It is here argued that IT departments should pay special attention to [2] security and backup management, communication network development, technical network management and maintenance, and user support arrangements [2, 3] (see Fig. 1).

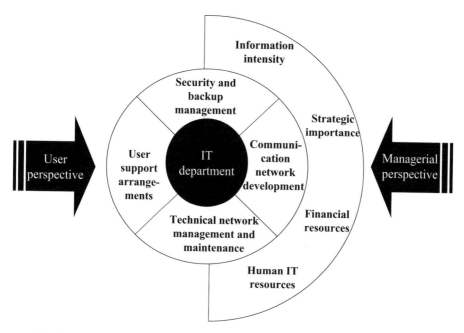

Fig. 1. Critical information systems management activities – addressing robustness.

3 Summary

Information systems infrastructure refers to a range of devices and technologies, applications and systems, standards and conventions that the individual user or the collective rely on to work on different organizational tasks and processes. Infrastructure is shared among different users, systems and processes, and it is intended for long time use [4]. It is the backbone of operations in the organization, and therefore it needs to be robust and trustworthy. Robustness is result of several activities. It is not only about selecting best possible technologies. More importantly, maintenance and development activities need to take robustness as a key priority, and make decisions that maximize reliability of systems. in addition, user awareness and support activities are important parts of successful information systems infrastructures. It is also important that infrastructures and systems remain open for changes, also flexibility is needed [5].

References

1. Urbach, N., Ahlemann, F.: Die Entwicklung der Unternehmens-IT – Von den Anfängen bis zur IT-Organisation der Zukunft. In: IT-Management im Zeitalter der Digitalisierung. Berlin, Heidelberg. Springer, Heidelberg, pp. 21–34 (2016)
2. Sirkemaa, S.: Infrastructure management: experiences from two case organizations. In: Proceedings of the 24th Information Systems Research Seminar in Scandinavia, Ulvik, Norway (2001)
3. Sirkemaa, S.: Towards information technology infrastructure management. J. Emerg. Trends Comput. Inf. Sci. 6(11), 614–621 (2015)
4. Hanseth, O.: Information Infrastructure Development: Cultivating the Installed Base. Studies in the Use of Information Technologies, No 16, Departments of Informatics, Goteborg University (1996)
5. Duncan, N.B.: Capturing flexibility of information technology infrastructure: a study of resource characteristics and their measure. J. Manag. Inf. Syst. 12(2), 37–57 (1995)

Standards and Information Systems Management – The Key to Success

Seppo Sirkemaa(✉)

University of Turku, Turku, Finland
seppo.sirkemaa@utu.fi

Abstract. Information systems management is expected to deliver systems and solutions for various processes and functions throughout the organization. Ultimately, the goal is to provide tools for users so that they can do their work. This calls for robustness in technologies and systems, for the information systems department the goal is to develop information systems that run smoothly and without unexpected interruptions. Standards and standards compliance is the key to developing robust infrastructures. In addition, systems that are built on standards are compatible to existing and future technologies. They are also needed to make information system flexible, and allow changes to be made rapidly and cost-effectively.

Keywords: Information systems · Infrastructure · Technology
Flexibility · Standards

1 Introduction

Information systems have an important role in organizations today, practically in all kinds of private and public organizations. Information systems have a key role in operations, in organizations operating in different fields and industries. Technology, especially information systems, computers and networks are so vital part of processes and activities that even small interruptions halt operations and can have a major impact on the whole organization. Furthermore, the importance of information technology on operations and profitability is ever increasing as more and more services are provided and becoming accessible over internet.

In recent years, the popularity of mobile technologies and solutions have rocketed. Using mobile devices like smartphones or tablets, has become the primary way to use and access information systems. This is clearly a challenge for developers of modern information systems and services. Systems and services should function smoothly also with smaller terminal devices, in all kinds of conditions ranging from bright daylight to total darkness, and in environments where connections are slow or unstable. At the same time, business is moving from 'brick and mortar' to internet, and the potential of electronic business seems enormous. Clearly, there is a need for systems and solutions that support business innovation to keep up with digital transformation in the economy [1].

In general, information systems management is a challenging task. Technology has such a critical role that the reliable and smooth operation of systems is a necessity for

© Springer International Publishing AG, part of Springer Nature 2018
T. Antipova and Á. Rocha (Eds.): MosITS 2017, AISC 724, pp. 245–251, 2018.
https://doi.org/10.1007/978-3-319-74980-8_23

practically any business activity and process. It is not uncommon that there are limited resources in information systems management department, especially in smaller organizations. It is still expected that information systems run with minimal interruptions, and function as a robust basis for company's activities and processes, whatever they might be now and in the future [2, 3].

Information systems should be robust, but also flexible for changes in the future. This means that there is a need for both robustness and flexibility in information systems. These can be contradictory requirements, as developing robust systems requires plenty of development work, planning and thorough testing, which tends to be a long and time-consuming process. Flexibility on the other hand is about creating new services and solutions rapidly, often on top of existing systems and technologies, and with minimal changes to other systems. Developing an information system that is robust and flexible - how can it be done?

We believe that selecting standard technologies and using standards compliance solutions is the key to robust infrastructures. In addition, systems that are built on standards are likely to fit with existing systems, and better compatible with emerging technologies. Standards are needed to create systems that can connect and interact, but also the way how data is being created and shared needs to be standardized.

2   Information Systems Development Activities

It is typical that information systems development is affected by several issues, both internal and external. Here external issues are referred to as changes in environment that cannot be influenced by organizations own actions, like competitor actions. This does not mean that nothing could be done. It is important to build flexibility into the systems and infrastructures so that adapting to changes are possible fast and with minimal resources. Another group of issues affecting information systems development have their roots within the organization. For example, changes in organization and the way resources are prioritized affects information systems requirements, and dictates information systems development. Similarly, setting organizational business goals and targets should also have an impact on information systems development activities. In addition, accessibility, reliability, and performance are aspects that need to be taken into consideration in information systems development [2–4].

Accessibility refers here to reach of the infrastructure, to areas and locations where user can access the system. In the best possible case access is possible with different devices ranging from desktop computers to variety of portable devices independently of their operating systems. Expressed in other words, if only certain geographical locations, terminal devices or operating systems can be used in accessing the information systems, this can severely affect the way people do their work. It also impacts the flexibility to adapt to changes in business environment. Therefore, developing an infrastructure that does not limit operations, is straightforward and seamless to the user maximizes the use of organizations resources.

Generally, reliability is considered among the most important features of information systems [2, 3]. It is expected that organization's information is 'up and running', available whenever needed, with minimal downtime. Information systems should also be responsive and fast. In addition, some environments list quality of services particularly important. For example, in systems that provide online services can transactions depend on interaction of several subsystems and modules. Increasing system usage and inadequate capacity causes problems, system is not responsive and interactions terminate unsuccessfully. This will lead to negative user experiences, unhappy customers, and indicate that there is need for further development, to make the systems and services more trustworthy [5].

Uninterrupted operations are the key to successful operations, therefore reliability, robustness and failsafe features need to be built into systems and infrastructures. This can however, be costly and therefore it is a question of resources, priorities and business requirements. Clearly, this requires planning and decisions need to take into consideration not only technical aspects, but also business needs. It is typical, that systems grow as more computers and users are connected to existing systems and infrastructures. Growth and expansion should be taken into consideration already in information systems planning, otherwise, an organization might outgrow it's infrastructure [4, 6]. Systems may be limited by number of simultaneous users, connected devices or other issues. These limitations can become barriers to expansion, it may turn out that adding capacity cannot be done incrementally, and there is need for significant investments. Consequently, choices that have been made earlier in systems planning can impact system expansion, costs involved, and may also influence reliability and quality related issues [7]. In information systems infrastructure investments and fixed costs can often be significant and therefore it is expected that technologies and systems will serve for years ahead. Therefore, making choices that guarantee compatibility, robustness and usability in the long run is a key development issue. Technologies also need to fit into changing business needs, and so they should be flexible and adaptable to different requirements, allowing a rapid provisioning of standardized resources [8].

Information systems development is not only a technological challenge. Choosing technologies and systems may present a unanimous picture to system designers. However, any organizational unit and any individual user confronted with new technologies and infrastructures have to find their own ways of integrating the technology into their work practices. The users' viewpoint to information systems development involves also the transition from old processes to new routines and usage patterns, and it is more diverse than IT developer's perspective. In the implementation of new technologies and systems it is possible that new technologies and functionalities may only be partially perceived and integrated into the users' practices. This underlines the fact that successful information systems development involves lot more than choosing best technologies.

3 Development and Standards

Information systems and infrastructures are typically combinations of different technologies and systems. Here, for independent information systems to collaborate to meet desired goals, they need to be able to communicate. Interoperability between systems can be achieved though communication. Standards have a key role in communication, and they affect development in several ways [9, 10]. Relying on manufacturer or vendor-specific technologies and systems may in some specific cases be effective. However, in a general-purpose infrastructure are widely accepted industry standards and practices necessary [11]. Technologies and systems that follow standards are often better than technologies that are not compatible or otherwise differ significantly from other technologies and solutions that have been installed and are in use in the organization. This is because there is often a need to connect systems together, allowing resource sharing or data integration for management purposes, for example.

Integration is often considered important because separate information systems make information sharing challenging. As an example, entering and maintaining customer data in separate systems is often not as effective as sharing common data through different systems and functions. This is, however, not possible without compatibility between different technologies and systems. This is where standards have a critical role. Standards make it possible to add new modules to existing systems and they make connecting technologies and systems from different providers possible. Sharing data between systems, and interoperability among different processes becomes possible when technologies, systems and their connections or interfaces are standardized, and the way they connect have been agreed on.

The concept of standard is central to the development of compatible and open information systems. Standards are agreements within the organization or between organizations where stakeholders conform to a set of principles and norms that guide information systems planning decisions. Standards can be considered as an effort to promote utilization of protocols and formats that enable applications developed by different software vendors to interoperate [12].

For organizations information systems development standards are about decisions that define the key systems and technologies which are required to connect and make different systems, modules or other technologies work together. Standards can facilitate collaboration, increase efficiency, reduce costs and complexity, and allow sharing of data and information [13]. A widely accepted, well defined standard is invisible to the user, but it enables technologies and systems that have been developed by different manufacturers to interoperate. Every organization needs to ensure that its information systems address standards, or else understand what might happen if standards are ignored [14]. Issues like integration, interoperability and standards compliance are the key drivers in information systems development. Therefore, instead of picking the best possible individual technology for a specific purpose companies need to increasingly choose the best system within a standard [4].

If we look at interoperability, technologies and infrastructures, we need to see beyond the technical, hardware part of information systems. The way how data can be made compatible and possible to transmit between different systems need to be agreed on, it is not about technical compatibility alone. Interoperability calls for standards between systems in what language, transmission method, collaboration protocol and the security will be used [13]. Standards can range from narrow to broad, specific to vague, complex to simple, or informal to formal [4]. Therefore, it is important that developers of information systems have a clear and shared understanding of the standards, procedures and activities which eventually will impact technological development choices. If there are multiple interpretations of a standard or other unclear issues concerning the system which is under development, it is likely that there may be incompatibilities, or lack of interoperability between systems that are being designed.

In general, using standardized technologies and solutions that follow standards makes sense. This can be explained by the self-reinforcing mechanisms of standards [9]. In information technology the self-reinforcing process refers to choosing systems that are compatible with existing technologies. When new technologies fit into older infrastructures, the value of the whole system increases, bringing benefits to existing system users (figure). The mechanism works in two ways: First, a large installed base tends to attract complementary technological choices, making the chosen standard cumulatively even more attractive. Secondly, larger technological base makes the standard more credible, and together they make the standard attractive and further increase the compliance of the installed base [10, 15–18]. As a result, both the cause and effect of standards compliance are connected to the existing infrastructures and systems (Fig. 1).

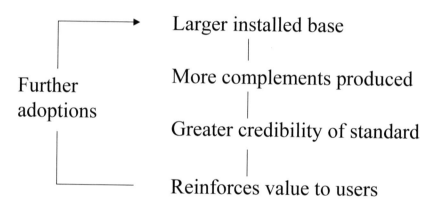

Fig. 1. Standards reinforcement mechanism

Developing systems that support a "standard" that builds on an installed base becomes more attractive than other possible solutions that do not fit the existing system. Other possible development options would lead to high costs in implementation or maintenance stage. Non-standard systems can require new knowledge to master, and so it may take a long time to obtain knowledge and skills for effective use of systems [9]. The self-reinforcing mechanism means that development becomes affected by the choices made in the early stages of the system development [17]. This mechanism is also referred to as the 'irreversibility of the installed base' [18]. Compatible technologies make it possible to copy technologies, systems and processes into similar environments (like implementing technologies and systems to offices across the country) and they allow tighter coordination in a common, integrated system infrastructure.

The compatibility of technologies impact development activities in several ways [9, 10]. Noteworthy, existing infrastructures, technologies and systems become a 'standard' which develop 'lock-in' to these technologies. This refers to a situation where existing, installed information systems infrastructure makes it impossible, very expensive or otherwise unwise to switch to other competing technologies. The situation leads to path-dependence where previous choices have an impact in the future, and make changes later difficult. Furthermore, the self-reinforcing nature of information systems development may also lead to inefficiencies because best solutions and technologies will not always be chosen due to compatibility issues.

Self-reinforcing mechanisms of standards help in understanding the choices made in development of information system. Here standards make it possible to develop systems that are compatible and integrated, but there are also other factors that support the adoption of technology that fits into the existing, installed base [19].

Having compatible infrastructures, technologies and systems that fit together are among the most important information systems development goals. Reliable, smoothly operating information systems are backbone of all operations and activities in the organization [2, 3]. Standards have significant importance because manufacturers develop technologies and services that adhere to standards, building over existing infrastructures and technologies. There is a market for compatible products, and they also call for support services and expertise in technologies. Clearly, the increasing number of users of a technology also leads to economies of scale and reduces prices, and this makes the technology even more attractive for the potential buyer. When a standard becomes widely accepted, manufacturers' compliance is almost guaranteed because it offers access to wider markets [4].

4 Conclusion

In information systems development is challenged to create infrastructures and systems that are robust and reliable. When systems and applications are being developed, new technology needs to fit and integrate into existing technologies and infrastructures. Infrastructures and systems also need to be flexible for future changes, in technologies and business demands.

Standards and standards compliant technologies are the key in creating flexible infrastructures. It is not wise to intentionally create incompatible systems that cannot connect or share data with other systems in the organization. Using technologies that are not compatible, or do not follow the standards adapted in the organization can be unwise and costly decision [18]. In this way development decisions that have been made in the past tend to impact development now and in the future, thus creating lock-in behaviour and path-dependence in information technology infrastructure management [17, 18]. However, using standard technologies and solutions helps in creating infrastructures that allow interconnecting systems from different providers, enable interoperability and makes infrastructure flexible for changes.

References

1. Urbach, N., Ahlemann, F.: Die Entwicklung der Unternehmens-IT – Von den Anfängen bis zur IT-Organisation der Zukunft. In: IT-Management im Zeitalter der Digitalisierung, pp. 21–34. Springer, Heidelberg (2016)
2. Sirkemaa, S.: Infrastructure management: experiences from two case organizations. In: Proceedings of the 24th Information Systems Research Seminar in Scandinavia, Ulvik, Norway (2001)
3. Sirkemaa, S.: Towards information technology infrastructure management. J. Emerg. Trends Comput. Inf. Sci. 6(11), 614–621 (2015)
4. Keen, P.G.W., Cummins, J.M.: Networks in Action: Business Choices and Telecommunications Decisions. Wadsworth Publishing Company, Belmont (1994)
5. Whitmore, J.J.: A method for designing secure solutions. IBM Syst. J. 40(3), 747–768 (2001)
6. Nadler, J., Guarnieri, D.: How to Keep Your Novell Network Alive: Survival and Success in a Multi-Vendor Environment. Bantam Books, New York (1993)
7. Nadig, D.V., Hard, N.J.: A proposed model for managing local area networks and measuring their effectiveness. In: Proceedings of the 26th Annual Hawaii International Conference on System Sciences, pp. 538–547 (1993)
8. Glohr, C., Kellermann, J., Dörnemann, H.: The IT Factory: A Vision of Standardization and Automation, The Road to a Modern IT Factory, pp. 101–109. Springer, Berlin (2014)
9. Arthur, B.W.: Competing technologies, increasing returns, and lock-in by historical events. Econ. J. 99, 116–131 (1989)
10. Shapiro, C., Varian, H.R.: Information Rules: A Strategic Guide to the Network Economy. Harvard Business School Press, Boston (1999)
11. Hanseth, O., Monteiro, E., Hatling, M.: Developing information infrastructure: the tension between standardization and flexibility. Sci. Technol. Hum. Values 21(4), 407–426 (1996)
12. Chiu, E.: EbXML Simplified a Guide to the New Standard for Global E-Commerce. John Wiley & Sons Inc., New York (2002)
13. Irimia, V.: Information and knowledge; communication patterns in standards and technologies for economic information systems interoperability. Acta Univ. Danub. 8(3), 54–61 (2012)
14. Darnton, G., Giacoletto, S.: Information and IT infrastructures. In: Information in the Enterprise: It's More Than Technology, Salem, Massachusetts, pp. 273–294 (1992)
15. Duncan, N.B.: The Invisible Weapon: A Study of Information Technology Infrastructure as a Strategic Resource in the Insurance industry. Texas A&M University, Department of Business Analysis and Research, TX (1995)
16. Duncan, N.B.: Capturing flexibility of information technology infrastructure: a study of resource characteristics and their measure. J. Manag. Inf. Syst. 12(2), 37–57 (1995)
17. Grindley, P.: Standards, Strategy, and Policy. Cases and Stories. Oxford University Press, New York (1995)
18. Hanseth, O.: Information Infrastructure Development: Cultivating the Installed Base. Studies in the Use of Information Technologies, No. 16, Departments of Informatics, Goteborg University (1996)
19. Katz, M.L., Shapiro, C.: Network externalities, competition and compatibility. Am. Econ. Rev. 75(3), 424–440 (1985)

Implementation of MANETs Routing Protocols in WLANs Environment: Issues and Prospects

Abraham Ayegba Alfa[1], Sanjay Misra[2(✉)] [iD], Adewole Adewumi[2],
Fati Oiza Salami[1], Rytis Maskeliūnas[3], and Robertas Damaševičius[3]

[1] Kogi State College of Education, Ankpa, Nigeria
abrahamsalfa@gmail.com, fati.salami@kscoeankpa.edu.ng
[2] Covenant University, Otta, Nigeria
{sanjay.misra, wole.adewumi}@covenantuniversity.edu.ng
[3] Kaunas University of Technology, Kaunas, Lithuania
{Rytis.maskeliunas, robertas.damasevicius}@ktu.lt

Abstract. In general, communication is the process of sending and receiving data packets between several participants or nodes connected within a network. There are two main ways of establishing communication. The first is the WLAN in which a defined network base is used to provide services to senders and receivers of data packets across wireless medium. The second is the MANETs that allow direct communication between senders and receivers nodes for the purpose exchanging of packet data through the wireless and baseless station in the network structure. The major challenges of WLANs are the cost of maintaining base station, network congestions, low throughput and delays. MANETs was considered appropriate for providing communication services in WLANs structure. This paper implemented AODV and DSR MANETs protocols in WLAN environment and analyzed its performance. The outcomes were used to proffer practical WLANs design recommendation of the MANETs routing protocols in this scenario.

Keywords: MANET · Routing protocols · AODV · DSR · WLANs
Communication · Networks · Packet exchange · Receiver · Sender

1 Introduction

Communication networks provide a number of advantages over systems in which a point-to-point line enables only two participants to communicate with each other [1, 2]. The use of wired networks is either been complemented or gradually replaced by wireless networks, optical, satellite and other media. The use of wireless networks has become more popular due to the rapid deployment capabilities which have encouraged many researchers around the world [3]. A wireless network comprises interconnections of different nodes that are established requiring no use of wires. Portability is the most significant property of wireless network offering minimal stationary cost [4]. These devices perform other important functions such as packets routing from source to destination, flow control maintenance and error control throughout the network.

© Springer International Publishing AG, part of Springer Nature 2018
T. Antipova and Á. Rocha (Eds.): MosITS 2017, AISC 724, pp. 252–260, 2018.
https://doi.org/10.1007/978-3-319-74980-8_24

Infrastructure-less networks can be subdivided based on the types of application namely; wireless mesh networks, wireless sensor networks, wireless local area networks, and wireless ad-hoc networks [4]. However, WLANs designs are ineffective because of the high cost of maintaining station bases, network congestions, low throughput and delays. This paper investigates the effectiveness of AODV and DSR MANETs routing protocols deployed for WLANs environments.

2 Communication Networks

Communication networks based on serial data transmission are the platform of up-to-date automation systems. Whether this is office automation or automation of manufacturing or process plants, the task remains always the same, exchanging data between different devices or participants within a system. Communication networks deliver a number of advantages over systems in which a point-to-point line allows only two participants to communicate with each other [5, 6].

Usually, the classification criterion for majority communication networks is the distance. These include local networks, LANs (Local Area Networks), as well as widely distributed networks, WANs (Wide Area Networks). With LAN, emphasis is on speed and powerful data exchange within a locally restricted zone, whereas WAN should be able to transmit data on diverse data media across several thousand kilometres [7].

The network topology refers to the physical organisation of the participants in relation to each other within a network. This refers to the logical organization of the participants in relation to each other for the purpose of communication regardless of type of same as the physical arrangement. In order to connection more than two participants, different networking structures are available as shown in Fig. 1 [8].

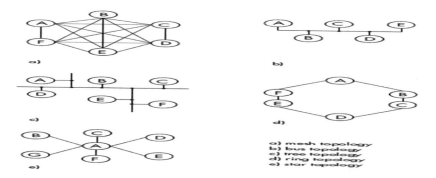

Fig. 1. Types of network topologies [8].

Complex network structures often consist of several, partly autonomous subnets. Each subnet can be based on a different topology and a different communications protocol as shown in Fig. 2. In any case, each participant must be able to clearly identify any of his communication partners and address him directly. In addition to this, any of the communication participants must be able to access the transmission medium. The transmission lines are assigned to the various participants by the protocol speci- fication that defines the access method [8].

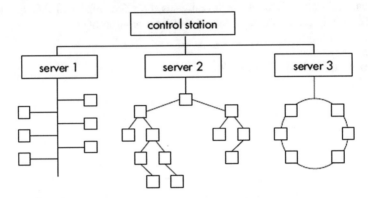

Fig. 2. A complex network structure of several subnets [8].

2.1   Wireless Local Area Networks

The wireless adapter (such as PCI cards for desktops or PC Cards for notebooks) is used to establish a wireless connection. A WLAN system is dissimilar to a traditional wired LAN in various ways including: (a) the destination address is not equivalent to a physical location, (b) WLANs deal with fixed, portable and mobile stations and, (c) the physical layers used here are fundamentally different from wired media [9].

In WLANs, the medium access control (MAC) protocol is the main component for determining the efficiency of sharing the limited communication bandwidth of the wireless channel. The fraction of channel bandwidth used by successfully transmitted messages gives a good indication of the protocol efficiency, and its maximum value is known as protocol capacity. IEEE 802.11 wireless network can be configured into two separate modes as Ad-Hoc and Infrastructure [9, 10].

WLANs offer easy wireless connectivity and access to viable network resources and services. Wireless networks are superior to wired networks because it is easy to install and flexible. Nonetheless, many wireless networks have features such as lower bandwidth, higher delays, higher bit-error rates, and higher operational costs in com- parison with wired networks. These applications have different demands from the core network protocol suite. Nowadays, high bandwidth internet connectivity stands tall as a basic requirement for the success of almost all of these networks structures [8, 9].

2.2   Mobile Ad-Hoc Networks

A wireless ad hoc network is a network without a specific base stations (that is, infrastructure-less or multi-hop); it is a collection of two or more devices connected together over wireless communications and networking capability to support computation anytime and anywhere. One type of this network is referred to as mobile ad-hoc network (MANETs) [11].

MANETs are autonomous entities that communicate through multiple wireless hops with the nodes acting as both clients and routers to forward data packets [12]. Benefits of MANETs include: on demand setup, fault tolerance, and unconstrained connectivity. Also, MANETs are being used in places where wired network and mobile access is neither productive nor feasible [13].

Dynamic Source Routing (DSR) protocol, Ad-hoc On-Demand Distance Vector (AODV) protocol, and Temporary Ordered Routing Algorithm (TORA) protocols fall under reactive routing approach [4]. A scheme designed to provide collision-free medium access to real-time traffic. The IEEE 802.11e standard contains an updated access mechanism referred to as the Enhanced Distributed Coordination Function (EDCF) with support for multiple traffic classes [14].

2.3 Transmission Control Protocol

Transmission Control Protocol (TCP) commenced as the Internet working solution allowing communication across a wide variety of media networks. Recently, TCP has been deployed over the Internet because it supports for diverse kinds of applications [15]. TCP is the standard networking protocol on the Internet widely used to transport protocol for data services such as File Transfer Protocol (FTP), Electronic Mail (e-mail) and Hyper Text Transfer Protocol (HTTP); and uses flow control, sequence number, acknowledgement and timer to ensure delivery from the sending process to the receiving process correctly, orderly and error-free [16].

These protocols can be categorized according to the routing strategy that they follow to find a path (route) to the destination. Routing protocols may be classified based on their routing table setup as proactive routing protocol, reactive routing protocol (on-demand) and hybrid as illustrated in Fig. 3.

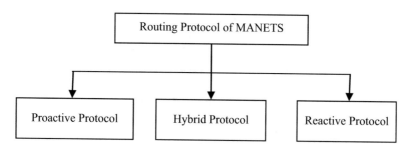

Fig. 3. Types of network routing protocols for MANETs.

3 Methodology

This paper adopted the OPNET 14.5 Modeler Version for the analysis of the two concepts understudied, which are WLANs and MANETs, both having few similarities and dissimilarities in terms of networks frameworks [17].

Mobile Ad hoc networks: The various entities such as application, profile, mobility, server, nodes are obtained from the object palette on the tool bar menu.

Application configuration is used traffic generated from each and processed on respective servers in the networks such as FTP, Email and HTTP (Delay, Network load and Throughput) [17].

Mobility configuration: The mobile nodes have a trajectory set to vector at 20 m/s with random waypoint mobility.

Server configuration: This module controls the traffic for MANETs protocol selection and TCP variant is selected through this module [17].

Mobile nodes: Nodes are work stations with client server applications running over TCP and UDP. The mobility options are set dynamically by allocating the IP addresses to all mobile nodes including server as well as the routing protocols [17].

Then, a wireless local area network (WLAN) server with applications running over TCP which support standard (IEEE 802.11) and its connection at 11 Mbps [17].

The global statistics is Wireless LAN that includes delay and throughput to determine the effectiveness of the AODV and DSR routing protocols [17].

The simulation steps and procedure are presented in Fig. 4.

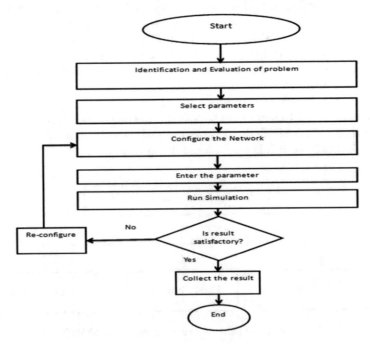

Fig. 4. Simulation steps and procedure.

3.1 Simulation Metrics

The parameters and variables for the OPNET 14.5 simulation analysis are contained in Table 1.

Table 1. Minimum parameters and values for simulation.

Metrics	Values
Performance Variables	Packet end-to-end delay, Network load, Throughput
Simulation time	130 s
Traffic type	FTP, Email and HTTP
Network size	1500 m × 1500 m
Number of nodes	50
Node speed (m/s)	20
Packet size	1 KB
Mobility model	Random waypoint mobility
Routing protocols	AODV, DSR
Data rate	10 Mpbs
Hardware	
RAM speed	1 GB
Hard disk capacity	120 GB
CPU speed	2.0 GHz

In Table 1, the packet-end-to-end delay measures the time when data reaches the MAC layer until successfully transmitted out of the wireless medium, which expected to be low to support real-time flows. Throughput measures the consumption of networks resources and wireless medium, which expected to be low for scare wireless bandwidth. The network nodes measure the capacity of network medium for increased load which can be impacted by the different number of nodes and load strengths. These factors largely affect the reliability of WLANs.

4 Results

Experimental outcomes of the three setups for the two MANETs routing protocols (AODV and DSR) using 50 network nodes in WLAN environment. In first setup, the WLAN delay load is measured for AODV and DSR routing protocols for 50 nodes is shown in Fig. 5.

In Fig. 5, the result of the simulation for the delay depicts that AODV and DSR begin propagation at the same time at 10 s. The AODV has the highest delay at 0.025 s (33.33%) after 10 s of the simulation and decreases steadily to the lowest delay at 0.08 s at the end of the simulation period of 130 s. While, DSR has the highest delay of 0.12 s (66.67%) after 20 s of the experiment and slides down like a curve shape to the its lowest delay of 0.06 s after 130 s or end of the simulation. This implies that for 50 nodes in the WLAN, AODV performs better than DSR in terms of WLAN delay, because the time taken for finding the path or route is lower for the AODV than the DSR.

In second setup, the WLAN network load is measured for AODV and DSR routing protocols for 50 nodes is shown in Fig. 6.

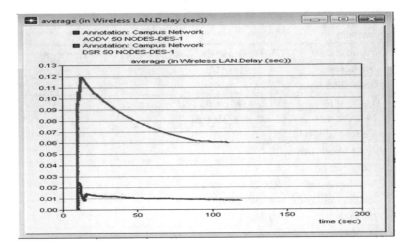

Fig. 5. Wireless LAN delay for AODV and DSR.

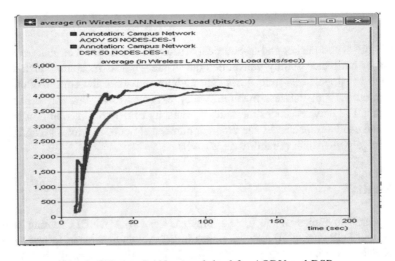

Fig. 6. Wireless LAN network load for AODV and DSR.

The entire network loads for the WLAN for 50 nodes are shown in Fig. 6. The AODV and DSR have almost the equal minimum network loads (100 bits/sec) at 10 s of the experiment. The network loads increases steadily from the start of the experiment to the finish. The difference between the maximum network loads is 200 bits/second for AODV and DSR routing protocols for the WLAN after 70 s and 120 s respectively (48.24% for DSR and 51.76% for AODV of the network load for the entire network load in the experiment). Therefore, DSR is better for WLAN as against AODV in terms of network loads transferred over 50 nodes in the period of 130 s.

The through of the wireless LAN is measured for AODV and DSR based on 50 nodes is shown in Fig. 7.

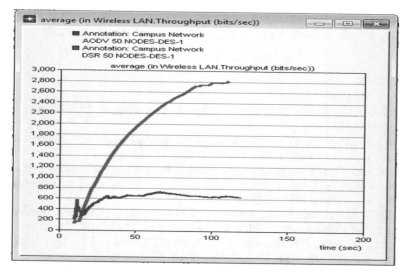

Fig. 7. Wireless LAN throughput for AODV and DSR.

In Fig. 7, the throughput of WLAN for the AODV and DSR are represented as minimum values at 10 s with 180 bits/second and 200 bits/seconds respectively. The throughput for DSR increases steadily until reaches the peak value of 2800 bits/seconds (28.87%) after 110 s of the experiment. While, in the case of AODV, the throughput increases up and down during the period of the simulation, then climaxed at 6900 bits/second (71.13%) after 130 s. These revealed clearly that the DSR out-performed AODV in WLAN environment.

5   Conclusion

The nature of WLANs is predominantly to control messages for large number of nodes by the use TCP usually between senders and receivers across base stations. The TCP can be problematic in MANETs which is investigated in this paper.

The paper revealed that, in the WLAN, AODV performed better than DSR in terms of WLAN delay and throughput, because the time taken for finding the path or route and bits/second of packets are lower for the AODV than the DSR. The reverse is the case for DSR, because it performed better in WLAN as against AODV in terms of network loads transferred over 50 nodes in the period for network load compared. There is greater efficiency when MANETs routing protocols are deployed for WLANs network designs especially for scalability and lower packets delays/losses. Consequent upon these, the paper recommends both MANETs routing protocols be deployed WLAN situation. Also, WLANs built upon MANETs routing protocols offer better performance through enhanced communication directly between different numbers of mobile nodes without the need for base stations as the case of traditional WLANs. However, there is need to conduct further tests using more network nodes and metrics.

Acknowledgement. We acknowledge the support and sponsorship provided by Covenant University through the Centre for Research, Innovation and Discovery (CUCRID).

References

1. Adewale, O.S., Falaki, S.O.: Internet Telephony Fundamentals. COAN Series, vol. 12, pp. 186–214 (2001)
2. Jain, S.: Performance analysis of voice over multiprotocol label switching communication networks with traffic engineering. IJARCCE **2**(7), 195–199 (2012)
3. Amar, N.M., Deepak, A., Shashwat, S., Vineet, S.: Performance evaluation of mobile ad-hoc network (MANET) routing protocols GRP, DSR and AODV based on packet size. Int. J. Eng. Sci. Technol. **4**(6), 2849–2852 (2012)
4. Harpreet, S.C., Ansari, M.I.H., Ashish, K., Prashant, S.V.: A survey of transmission C.P over mobile ad-hoc networks. Int. J. Sci. Technol. Res. **1**(4), 146–150 (2012)
5. Gupta, J.S., Grewal, G.S.: Performance evaluation of IEEE 802.11 MAC layer in Supporting delay sensitive services. Int. J. Wirel. Mob. Netw. **10**, 42–53 (2010)
6. SAMSON: Communication Networks. Technical Information. Part I, L155EN, Frankfurt, Germany, pp. 1–31 (2012)
7. Robert, F.: Wireless Communications. Microsoft Encarta (2007)
8. Ahmed, M., Fadeel, G.A., Ibrahim, I.: Differentiation between different traffic categories using multi-level of priority in DCF-WLAN. In: Advanced International Conference on Telecommunications, vol. 6, pp. 263–268 (2010)
9. Puschita, E., Palade, T., Pitic, R.: Wireless LAN medium access techniques QoS perspective. In: Signals, Circuits and Systems, pp. 267–270 (2005)
10. IEEE Std. 802.11b: Wireless LAN Medium Access Control (MAC) and Physical Layer (PHY) Specifications: Higher-Speed Physical Layer Extension in the 2.4 GHz Band (Supplement to Part 11) (1999)
11. Amtabh, M., Baruch, A.: Introduction to ad hoc networks. CS-647 advanced topics in wireless networks, pp. 1–53. Department of Computer Science, John Hopkins University (2008)
12. Natarajam, K., Mahadevan, G.: A comparative analysis and performance evaluation of transmission control protocol (TCP) over mobile ad-hoc network (MANET) routing protocols. Int. J. Adv. Comput. Eng. Archit. **1**(1), 47–62 (2011)
13. Balder, R.M., Barwan, N.C.: Effect of mobility on performance of MANET routing protocols under different traffic patterns, pp. 19–24. IP multi-media communications, Jai Narain Vyas University (2005)
14. Sobrinho, J.L., Krishnakumar, A.S.: Quality-Of-Service in ad hoc carrier sense multiple access networks. IEEE J. Sel. Areas Commun. **17**, 1353–1368 (1999)
15. Papanastasiou, S., Mackenzie, L.M., Ould-Khaoua, M., Vassilis, C.: On the interaction of transmission control (TCP) and routing protocols in mobile ad-hoc networks (MANET). Department of Computing Science, University of Glasgow, United Kingdom (2006)
16. Muhammad, A., Muhammad, A., Zahid, U.K., Abdulsalam, A., Shazad, R.: Comparison of optimized link state routing (OLSR) and dynamic in-demand routing protocol (DYMO) on the basis of different performance metrics in mobile Ad-hoc networks. Am. J. Sci. Res. Inst. Manag. Sci. Peshawar Pak. **2**(2), 67–85 (2011)
17. OPNET: Version 14.5. Ljiijana Trajkovic: OPNET technologies. Inc. Bethesda, Maryland, United State of America (2008)

Author Index

© Springer International Publishing AG, part of Springer Nature 2018
T. Antipova and Á. Rocha (Eds.): MosITS 2017, AISC 724, p. 261, 2018.
https://doi.org/10.1007/978-3-319-74980-8

Printed in the United States
By Bookmasters